TEACHING FOR COMPLEX SYSTEMS THINKING

TEACHING FOR COMPLEX SYSTEMS THINKING

Rosemary Hipkins

NZCER PRESS

NZCER PRESS
Rangahau Mātauranga o Aotearoa | New Zealand Council for Educational Research
Te Pakokori, Level 4, 10 Brandon St
Wellington
New Zealand

www.nzcer.org.nz

© Rosemary Hipkins, 2021

ISBN: 978-1-99-004020-7

No part of the publication may be copied, stored, or communicated in any form by any means (paper or digital), including recording or storing in an electronic retrieval system without the written permission of the publisher.
Education institutions that hold a current licence with Copyright Licensing New Zealand may copy from this book in strict accordance with the terms of the CLNZ Licence.
A catalogue record for this book is available from the National Library of New Zealand.

Designed by Smartwork Creative Ltd

Printed by About Print, Wellington

Cover image
Sarah Slavick, *Elegy to the Underground* (2020).

Contents

Foreword — 1

1. Introduction — 3
   How the book came together — 5
   What does the book cover? — 5
   An outline of the structure of the book — 6

2. Complex systems thinking is important for every student — 9
   Example 1: Exploring the impact of choices and actions — 9
   Solving word problems in mathematics — 12
   Making complex systems thinking a specific learning focus — 14
   Working out how best to manage the COVID-19 pandemic — 17
   A quick note about scope — 20
   Chapter 2 reading guide — 22

3. Features of complex systems — 24
   Definitions that trace the development of the field — 25
   A summary of complex systems concepts — 26
   Why explicit teaching of this knowledge is needed — 35
   What might "explicit teaching" look like? — 36
   Chapter 3 reading guide — 37

4. Traditional teaching practices can undermine complexity thinking — 38
   Working on our own causal reasoning — 43
   Is change even possible? — 45
   Chapter 4 reading guide — 46

5. Pedagogies that support the development of complexity thinking — 48
   Keeping a dual focus on parts and wholes — 48
   Using modelling as an inquiry process — 50
   Focusing on processes/mechanisms, not just structures — 51
   Collaborative conversations and metacognitive processing — 52
   The *situated* nature of shared metacognitive conversations — 55
   Leveraging complexity ideas to create new pedagogies — 56
   Looking ahead — 58
   Chapter 5 reading guide — 59

6. Opportunities provided by e-learning resources — 61
   Introducing the "semi-quantitative" features of e-resources — 62
   Top-down models — 63
   Rethinking possibilities for an existing resource — 67
   Agent-based (bottom up) models — 68
   Finding learning resources that will work for you — 72
   Situating students *inside* rich learning experiences — 74
   Chapter 6 reading guide — 75

7. Situating ourselves *inside* systems	76
Introducing the idea of compassionate systems thinking	77
Pedagogies that bring thinking and sensing together	79
Some e-resources can immerse students inside systems	82
A few thoughts on working with models	86
Looking ahead	88
Chapter 7 reading guide	89
8. Learning from indigenous knowledge systems	91
Parallels between indigenous knowledge and complexity	92
Webs of relationships decentre individuals within the system	93
The metaphor of two-eyed seeing	101
A thought experiment in decentring ourselves	103
Chapter 8 reading guide	105
9. Fostering thinking dispositions by working on habits	107
A focus on dispositions and habits	108
Building habits for complex systems thinking	111
Building a varied repertoire of habit-building contexts	117
Chapter 9 reading guide	120
10. Curriculum-design considerations	121
Five clear curriculum messages	121
Translating good intentions into specific curriculum design	124
How subjects might contribute to complex systems thinking	126
Integration as a both/and design challenge	134
Collaborative, co-ordinated curriculum design	137
Paying attention to the taken-for-granted	138
Chapter 10 reading guide	139
11. Choosing contexts that support the development of complexity thinking	141
An exploration of generative types of contexts	144
Challenges to consider when choosing a mix of contexts for a programme of learning	153
Being clear about the overall learning focus	154
Chapter 11 reading guide	156
12. Innovative approaches to assessment	158
Assessing gains in knowledge of complex systems	160
Demonstrating awareness of interconnectedness	163
Designing problem-based assessment tasks	167
Evidence that students are building systems-thinking habits	169
Organising a portfolio of evidence	171
Being clear about assessment purposes	173
Chapter 12 reading guide	174

13. Indications that students are making progress	175
Progression is a complex phenomenon	176
Different models of progression (and how they might fit together)	178
A practical strategy for noticing and documenting progress	186
The complexities of meaning-making	188
Chapter 13 reading guide	190
14. Cognition is complex	192
Understanding the brain as a complex adaptive system	193
Constructivism as active meaning-making	195
Building awareness of changes in our own meaning-making	202
Exploring time as a complex concept	204
Chapter 14 reading guide	207
15. Complexity at work in the world	208
Complexity thinking is important for teachers	209
Complexity thinking underpins many management strategies	211
Complexity dynamics can be used to check claims to truthfulness	213
Complexity thinking invites the "unthinkable"	216
Once more with feeling: In conclusion	220
Chapter 15 reading guide	222
References	224
Acknowledgements	236
Index	238

Tables

Table 1. Beliefs about how the world is (slightly modified from Yoon, 2008)	43
Table 2. Ten key differences between general intelligence and adaptive intelligence problems (Sternberg, 2020, n.p.)	169
Table 3. A proposed developmental sequence for understanding concepts that explain systems dynamics (after Yoon et al., 2019)	180
Table 4. Potential indicators that students have made progress in building their conceptual and contextual knowledge of a complex system	183
Table 5. A synthesis of ideas about skills progression (after Plate and Munroe, 2014)	186
Table 6. Three different types of knowledge structures that our brains construct (after Lemmer et al., 2020)	198

Figures

Figure 1.	An electronic display of real-time electricity and water consumption in a school. (Source: Clark et al., 2017)	11
Figure 2.	A mathematical problem that needs to be addressed using systems thinking (Source: Salado et al., 2019)	13
Figure 3.	An example of secondary-school students' systems thinking (Source: Heinrich & Kupers, 2019)	15
Figure 4.	A causal-loop diagram to inform policy-making in response to the COVID-19 pandemic (Source: Bradley et al., 2020)	18
Figure 5.	The ladder of inference reflection tool (After Argyris, 1970; Senge, 1990)	54
Figure 6.	An example of a stock-and-flow simulation in SageModeler	64
Figure 7.	A game that could be repurposed as a systems-thinking resource	67
Figure 8.	Netlogo simulation of an aquarium system (Source: Hmelo-Silver et al., 2017)	70
Figure 9.	The EN-ROADS climate simulator (https://en-roads.climateinteractive.org/scenario.html?v=2.7.19)	84
Figure 10.	Using a causal-loop model to build a local curriculum centred on indigenous knowledge. (Source: Heke et al., 2019, p. 26)	99
Figure 11.	The doughnut model reimagined from an indigenous knowledge perspective (Source: Shareef, 2020)	100
Figure 12.	An example of one "Habits of a systems thinker" card, as developed by The Waters Foundation	113
Figure 13.	Habits of Mind, as developed by the Habits of Mind Institute	116
Figure 14.	An example of a simple inquiry into fractal patterns in mathematics	130
Figure 15.	An example of a systems drawing scaffold (Source: ARBS)	161
Figure 16.	A schematic overview of progression in complex systems thinking	188
Figure 17.	The basic arrangement of an "energy transfer model" (Nordine et al., 2018, p. 187)	199
Figure 18.	A summary model of learning-to-learn as a complex process (After Deakin Crick, 2014, p. 16)	205

Foreword

When Rose Hipkins and I met in person in a café in the Wellington CBD to discuss her early stage manuscript of this book, it was everything but a given that we would be able to do so. By that time, I had known Rose for a bit more than 2 years. In 2018 she was the responsible researcher at NZCER to oversee a research project funded under the Teaching and Learning Research Initiative (TLRI) scheme, which I co-led together with my colleague from education, Dr Dayle Anderson. Our interdisciplinary team was looking at how we could improve primary-school science education by involving online citizen-science participation in class activities. As a computer scientist with an interest in emergent phenomena, I was looking at aspects that relate to what happens when we put a particular piece of digital technology (i.e., the online citizen-science interfaces) in front of a particular group of people (i.e., primary-school children).

Now it was October 2020, quite a while after the end of the aforementioned research project, and the world as we knew it was in the strong grip of the COVID-19 pandemic. People in many countries, including my family in my birth city Berlin in Germany, were in enforced "lockdown", one of the most common public health interventions to stop the spread of the disease COVID-19 caused by the novel coronavirus known as SARS-CoV-2. Yet here in Aotearoa we were at that time free to live our life almost normally. Beside an increased awareness of adequate hand and cough hygiene as well as large-scale movement tracking, using a public health app on smartphones, it was often only the news from overseas that would remind people that we were still living in the middle of a global pandemic.

Rose and I were discussing what I think have become two key arguments at the heart of her book: complexity science doesn't make the world more complex than it is, the world is complex by nature. And understanding complexity is crucial within the contemporary social fabric, a fact that is not obvious to everyone, but that comes clearer than ever before when considering the COVID-19 pandemic again.

The power of applying a complexity-analysis framework in this context can be understood when looking at how models of disease outbreak characteristics were used for the design and implementation of the effective public health interventions that allowed Aotearoa and other countries to prevent or stop the uncontrolled spread of COVID-19 within communities. On the other hand, the natural complexity in our world and the mathematical constructs created to understand it (e.g., via human-contact networks or exponential-growth functions) may become incomprehensible to some, and give rise to misinformation that is assumed to provide easier explanations that allow people to regain control over an overwhelmingly complex and often unpredictable situation.

Ever since we first met, Rose Hipkins has stood out to me as one of the great promotors, critics, and explainers of education at all levels. With this book she does something extraordinary for Aotearoa. She advances our understanding not just of why and how complexity should be taught in schools, she also makes a unique case for why teaching practice will improve when people understand it from a holistic complexity perspective.

Complex systems carry the characteristic property of emergence, making it impossible to predict when and how our world is going to change. The only thing that is certain is that our world will change. Rose Hipkins' book transports this message, which may sound disappointingly hopeless to some. But the book also provides a powerful set of examples and explanations as to why and how teaching complexity theory and complexity thinking at school level will lead to the development of a generation of thinkers who can intellectually cope with the inherent uncertainty of the world and ultimately make society more resilient. By reading this book, people can go onto this intellectual journey towards an understanding and appreciation of complexity, and they will be empowered to help others to get onto it as well.

Dr Markus Luczak-Roesch
Associate Professor at Victoria University of Wellington
Associate Investigator at Te Pūnaha Matatini –
CoRE on Complex Systems and Networks

Chapter 1
Introduction

Five or so years ago I led an exploratory project focused on the addition of key competencies to *The New Zealand Curriculum* (Ministry of Education, 2007) (*NZC*). Our aim was to document ways in which weaving key competencies with more traditional content could potentially change the nature of students' learning experiences, making them more richly purposeful. My colleagues and I approached the challenge from a contextual perspective—the starting point for our analysis was provided by complex issues that we could see students would face in their futures. We explored issues and ideas related to: globalisation; consumerism; wellbeing; inequality and social justice; climate change; food security; and changing teaching practices as a complex challenge in its own right. Our findings were published in a small book called *Key Competencies for the Future* (Hipkins et al., 2014). The book culminated with a manifesto of sorts. The final chapter listed seven capabilities that we thought all students need to take into their lives as citizens, and which they therefore needed to develop across their years at school. One of these capabilities reads as follows:

> [Our aspiration is for young people who can] **look beyond immediate causes to consider the joined up nature of things and events in the world.** This sensitivity to interconnectedness leads to more thoughtfulness about the impact of personal and collective actions on others and on environments—local and more distant—and ultimately on the planet as a whole. It is underpinned by an ever-growing knowledge about how things work in the world (Hipkins et al., 2014, p. 136, emphasis in the original).

Things and events in the world have always been joined up of course. But with rapid increases in computing power, scientists have recently come to understand much more about the nature of the interconnectedness of everything—the world is much more "intertwingly" than we ever could have imagined in the past (Weinberger, 2012). These discoveries continue apace. Also, aspects of the way we now live have intensified interconnections and made our impacts on the planet much more evident. Events that are happening now underscore the imperative to ensure that all students have the opportunity to learn about complexity, and to develop their capabilities for complex systems thinking.

Complex dynamics are in play all around us all the time. The COVID-19 pandemic, which is raging as I am writing, is just one example. Climate change is another. No less dramatically, our bodies are made up of many complex interacting systems and we need to understand this, at least in principle, in order to manage our own wellbeing. This very brief snapshot only scratches the surface of how and why complex systems impact on our lives and the health of our environments, and indeed the planet as a whole.

This book is a sort of sequel to *Key Competencies for the Future*. With the *NZC* framework as the starting point, *Key Competencies for the Future* included a range of descriptions of innovative teaching and learning practices used by teachers working at the cutting edge of curriculum-design thinking. In this book I explore the "something more" (than current innovation) that complex systems thinking now demands. I aim to show how and why new types of pedagogies are needed, or at the very least the best of traditional pedagogies need to be adapted in new ways. Since complex systems thinking is applicable in a wide range of contexts, new pedagogies will need to be deployed across the full breadth of the curriculum.

Many of the research papers I have been reading endorse the recommendation from *Key Competencies for the Future* cited above. In a wide range of educational contexts, from early childhood (ECE) to tertiary settings, researchers are saying that it is important for students to develop an understanding of complexity, and the ability to think in terms of complex systems. However, as I found out, assembling practical advice on how to achieve these good intentions is not so straightforward. I hope I have done justice to the challenge, at least as it applies to the school sector, and at least for now. This has been an ongoing personal learning journey for me, and I anticipate that it is likely to be so for anyone raised within a Western education system.

How the book came together

This book began life as a review of research with a focus on teaching students how to become systems thinkers, and to understand how complex phenomena might behave. This literature review was not undertaken by following formal processes that are typical of reviews that seek to make new knowledge claims. Several recent reviews of that sort are included in the papers cited (in particular Yoon et al., 2018). Instead, as I read the papers located for me by NZCER's librarian, and by my own additional searching, I endeavoured to glean practical insights that might be useful in supporting teachers who want to become more deliberate in their teaching for complexity.

As the work unfolded I also began to look for sources other than formal research papers. Some important gaps were becoming apparent and I felt a strong need to try and address them. Specifically, I had not been successful in finding much advice about how to position students *inside* complex systems, even though the importance of doing this had been signalled in several papers. Indeed, I had said this myself at a recent symposium (Hipkins, 2019a). About halfway through the writing process I began to share specific chapters and dilemmas with colleagues around the world who I thought might bring fresh perspectives to expand my own thinking. Chapter 7 marks the pivot point where I began to feel my way forward in this space of more challenging personal learning, where there is less formal research to draw on (or maybe to hide behind).

What does the book cover?

Each chapter is based on an aspect of teachers' professional work where there are important issues to be addressed. Often the ideas and evidence presented in the research papers contributed indirectly to the insights and advice I have distilled. For example, useful assessment advice might be gleaned from a description of the evidence gathered to answer a research question about whether students had successfully developed named aspects of systems thinking.

I had the following practical questions in mind as I read the research details and drew inferences from common themes that emerged.

Conceptual questions: What features of complex systems are important for school students to learn about and why?

Pedagogical questions: Are there specific challenges in teaching for systems thinking? If yes, what advice is given about how best to address these? Are there research-based resources that schools can readily access?

Curriculum questions: What sorts of outcomes are envisaged for including systems thinking in the curriculum? At what stage of their schooling could students be introduced to systems-thinking practices? How might systems thinking best be included in the overall curriculum?

Assessment questions: (How) have systems-thinking capabilities been assessed? What might "progress" look like? What issues have arisen for assessment and reporting?

An outline of the structure of the book

Determining the most helpful order for the chapters was challenging. Complex challenges have many facets and I often found myself needing to hold off on an idea until more pieces of the bigger picture were in place. I came to think of the structure as a sort of spiral, which is itself a complexity motif (Brown, 2019). I have provided forward-signals when I introduce an idea that will gain additional layers in a subsequent chapter.

- Chapter 2 briefly illustrates the scope of the book using diverse examples, chosen because they illuminate different aspects of the teaching and learning issues discussed.

- Knowledge about complex systems does not always make common sense. Many concepts need to be explicitly taught. Chapter 3 introduces these.

- Chapters 4 and 5 address the pedagogical questions. I had planned to write only one chapter, but found I needed to discuss challenges for changing established pedagogies before I could explore more effective approaches.

- A significant amount of research and development effort is being put into the development of e-learning resources that provide interactive experiences to support students' growing understanding of complexity dynamics. Chapters 6 and 7 outline opportunities in this area.
- As already signalled, I became aware of the need to address the challenge of helping students to position themselves inside systems, when predominant ways of thinking place humans outside natural systems looking in. Indigenous ways of knowing are bound up in this challenge. Chapters 7 and 8 provide my provisional exploration of this ongoing learning space.
- When I began work on the curriculum and assessment implications, discussed in chapters 10 to 12, I came to the uncomfortable realisation that I had not given sufficient attention to the complexities of developing the dispositional aspects of key competencies. I tackle that challenge in Chapter 9.
- Two more uncomfortable insights led to the insertion of additional chapters. Quite late in the writing process I realised that I had also been taking contexts for granted, and that an explicit discussion of how to choose fruitful contexts for learning was warranted. Chapter 11 addresses this topic.
- How should we think about progression in complex systems thinking when progressions in any area typically traverse a trajectory from simple to complex? This is the conundrum I tackle in Chapter 13.
- Some papers focus on teaching implications when learning itself is understood as a complex phenomenon. Chapter 14 addresses this challenge.
- I originally envisaged that Chapter 15 would explore career options that require knowledge of complexity. The chapter does do this but also goes much further. It ends with a discussion of "unthinkable" questions and challenges now being opened up by research at the cutting edge of complexity thinking. Profound implications for all our futures suggest that building dispositions for complex systems thinking should be an entitlement for every school student.

It is important to note the limitations of this book. There is a substantial body of literature that focuses the complex nature of change in schools and education systems and indeed of leadership in education generally. These management ideas are beyond the scope of my work here, which is focused on teaching and learning.

The next chapter provides a broad overview of why learning about and with complex systems matters. Four brief descriptions of quite different types of research are provided. In each case, systems thinking or complexity thinking has helped unlock important insights, actions, and/or learning dispositions.

A brief note about the reading guides

A teacher who read the almost completed book noted that there was so much to think about that is was easy to lose track of the pieces. This teacher thought that a reading guide for each chapter might help. It would provide a focus for groups of teachers who wanted to discuss ideas together, or a basis for solo reflection. I was initially reluctant, having seen examples of reading guides that I personally considered added little to the actual text. However I took the plunge and was soon glad that I had. I enlisted a critical friend to make sure that I kept the interests of teachers foremost in my mind as I worked to shape questions that traverse the broad themes of each chapter, not just pieces of detail. In this way, I have tried to model the part/whole thinking which is integral to teaching for complexity. I have also endeavoured to honour the experience and expertise that teacher-readers will bring to the ideas presented. I don't need to tell you that teaching is so much more complex than many people will ever know.

Chapter 2

Complex systems thinking is important for every student

Complex systems are everywhere: systems inside systems, from the largest to the smallest of scales. The more researchers explore, the more complexities they find. Understanding what complex systems are and how they impact on our lives is important knowledge for all learners to gain over the course of their schooling. It is also important that students build dispositions to think in systems terms—doing this has both academic and citizenship benefits.

This chapter provides four initial examples of the types of phenomena that are in focus throughout the book. Each of these examples summarises a specific piece of research. I then briefly discuss some of the challenges raised by, or inferred from, the work in question. My aim is build an introductory snapshot of the importance of helping all students to learn to think in complex systems terms.

Example 1: Exploring the impact of choices and actions
The first example describes research undertaken with primary-school-age students. It was part of a bigger project that also worked with adults in the American community of Oberlin (there is more detail about the work

with adults in chapters 7 and 12). The research team described systems thinking as "a way of perceiving the world" that:

- recognises systems as being comprised of and exhibiting properties that result from dynamic interacting parts
- incorporates a refined understanding of levels of cause and effect, including indirect as well as direct consequences and relationships between parts
- situates the systems thinker within the systems they are studying (Clark et al., 2017, p. 2).

Some key features of complex systems are identified here. Chapter 3 expands on these. Note especially the third bullet point—we humans are embedded in complex systems. We cannot live outside of them. Traditional Western ways of schooling don't necessarily mean to imply that we sit outside and separate from the more-than-human world, but that impression is typically conveyed by industrial-age pedagogies (see Chapter 4) and by the ways Western knowledge is typically structured (see Chapter 8). By contrast, the ways of knowing and teaching used in indigenous cultures don't separate "mind from body, self from other, and human from the more than human world" (Davis et al., 2015, p. 230).

What did this research team do to address the challenges they outlined? Figure 1 shows an electronic screen board[1] that was installed in a school corridor to serve as a visual reminder of trends in the school's electricity and water consumption. In class, students learnt about systems involved in production and use of water and electricity. They then explored potential impacts of behavioural changes they might make to reduce their water and electricity use. They could then observe the real-time impact of changes they made in response to their learning (Clark et al., 2017).

---

1   The technology is made available to support community action. See https://www.environmentaldashboard.org/

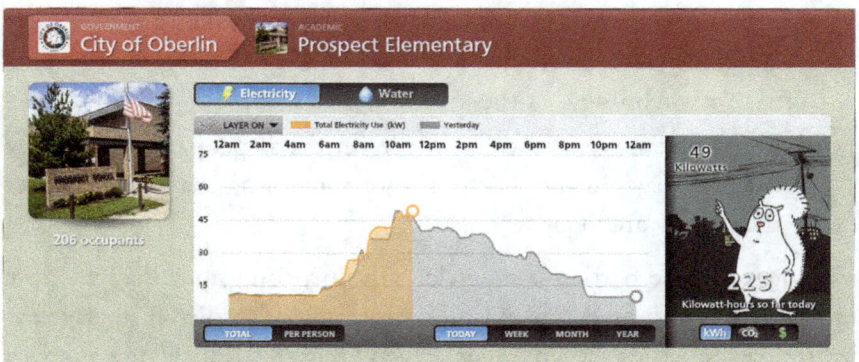

Figure 1. An electronic display of real-time electricity and water consumption in a school. (Source: Clark et al., 2017)

This research had a dual focus on learning about specific complex systems and directly observing consequences of personal and collective choices of consumption-related behaviours. The researchers hoped the experience would lead to increased self-efficacy and group efficacy for addressing conservation of water and electricity in the Oberlin community. They contrasted the learning outcomes for these students with those of a control group who experienced traditional learning about electricity and water consumption. Students in classes that used the display-board feedback showed greater content retention and content-specific systems-thinking skills than students in control classes. They also had significantly higher self and group perceptions of efficacy to act to conserve water and electricity. However, there was no difference between this group and control group in their *general* systems-thinking skills. We will come to why this might be in the next chapter.

Unpacking the challenges illustrated by this example

An Italian science-education research team reviewed a range of social-science research papers to suggest three types of reasons that people might resist taking action on complex issues such as climate change (Tasquier et al., 2014). They proposed three main types of barriers to personal involvement.

- The issue is seen as being too big for personal action to have any real impact—those in positions of power and authority should be acting.

- Lack of knowledge about complex issues can mean that risks appear small or irrelevant to the individual; or alternatively to be so overwhelming that the individual feels helpless.
- The problem seems too far in the future. If people perceive they need to "pay" in the short term for something that won't impact them, they are likely to resist.

None of these barriers are particularly surprising, but they do have implications for how complex issues are taught. Tasquier et al. say that learning experiences should: focus on ways that individuals interact with natural systems; explore causal connections and the dynamics that can cause small changes to have large impacts; and include timescales that support students to reflect on possible future scenarios (such as melting ice in the context of climate change).

Without greater detail, it is hard to know to what extent each of these three recommendations was addressed by the primary teachers in the Oberlin study. What is clear is that students had opportunities to gain real-time feedback about the consequences of personal actions, so that they could see how small changes, made collectively, can positively impact a system. Chapter 7 explores the important challenge of helping students to see themselves inside systems and therefore directly implicated in how those systems change over time. The way in which small changes can lead to much large impacts is an example of what is meant by something being "non-linear". This will be a recurring idea throughout the book.

Solving word problems in mathematics

My second example is set in the middle-school years (Salado et al., 2019). It is a small study: just three students were involved. I chose it because it surfaces an important issue about what is sometimes called the "hidden curriculum" of the classroom. How teachers lead the intended learning sends powerful signals to students about what is important.

Here's the pedagogical dilemma. When students are given word problems in maths, they are often encouraged to find and convert the numerical information to an equation to solve. But they can only do this quickly if they ignore contextual information provided for the problem. When teachers encourage them to do that, they essentially signal that contextual detail is clutter to be ignored. The example in Figure 2 was used by the researchers to discuss a different approach to word problems—one where asking questions about the contextual information is encouraged. This small tweak to the teaching process can change the way students think about the whole system in which the action takes place.

Figure 2. A mathematical problem that needs to be addressed using systems thinking (Source: Salado et al., 2019)

Students were asked about the time a family would need to leave the beach to arrive at their hotel at a specified time. It is actually impossible to say for sure because there are so many variables they might encounter along the way. The researchers worked with three students. The most able student mathematically (who, at 12 years old, was the also youngest) quickly solved the problem and was confident of her answer. She basically ignored all the contextual clues in the sketch. The student judged to be least able was age 13. He was the one who most clearly saw the difficulties and asked the most questions about unknowns. The researchers noted that his weaker computational skills would probably rule him out of even considering a career such as engineering, whereas they could see that he was displaying the sort of disposition to think in systems terms that all good engineers need. For students like this one, systems thinking might provide a more productive entry point to the computational aspects of mathematics (Salado et al., 2019).

Note that this particular paper talks about "systems thinking" not "complex systems thinking". There is a difference, which will be explained in Chapter 3.

A short reflection on the importance of mathematics

I chose this example to provide an early signal about the importance of question asking when exploring complex systems as messy, somewhat unpredictable wholes. Uncertainty is a challenge that teachers will inevitably face when exploring complex systems with students. When there cannot be one "right" answer, strategies are needed to deal with not knowing for sure. Teachers and students need to build dispositions to tolerate uncertainty, understanding that it is inevitable.

Markus Luczak-Roesch saw something else very important in this example. He is a complexity scientist. You've already met him briefly because he wrote the foreword, and I'll introduce some of his work in Chapter 15. For Markus, the most important aspect of this example is the spotlight it shines on school mathematics as the foundation on which research of complex systems builds. How mathematics is taught can encourage the disposition to think in complexity terms but must also lay the conceptual foundations for understanding number patterns, ratios, vectors, exponential change and, for older students, linear algebra. These are the working tools of many complexity scientists.

Making complex systems thinking a specific learning focus

My next example is set in an international secondary school in India. The researchers wanted to explore the impact of a unit of work that focused on complex systems thinking itself, rather than a traditional academic topic. They designed a series of seven sessions, each 2 hours long, that would gradually introduce students to the features and behaviours of complex systems (Heinrich & Kupers, 2019). The unit purposefully used different curriculum subjects as contexts for the aspects of complex systems being explored in each lesson. The researchers hoped that the students would recognise how they could take this new way of thinking about the world into their more traditional academic learning.

Figure 3 below shows the response of a small group of students to an end-of-unit exercise. They were given an outline of a cityscape and asked to annotate it to address two questions: what would a good life in a city be like? What would have to change? The researchers were looking for evidence of what they called "knock-on effects" from the unit, which might show up as:

- actively looking for interconnections when solving problems
- displaying a higher sensitivity for systems as wholes
- asking different types of questions
- valuing discussion of complexity models in subject lessons
- finding connections between learning contexts and daily events
- critically engaging their own assumptions and worldviews
- gaining new perspectives on societal challenges.

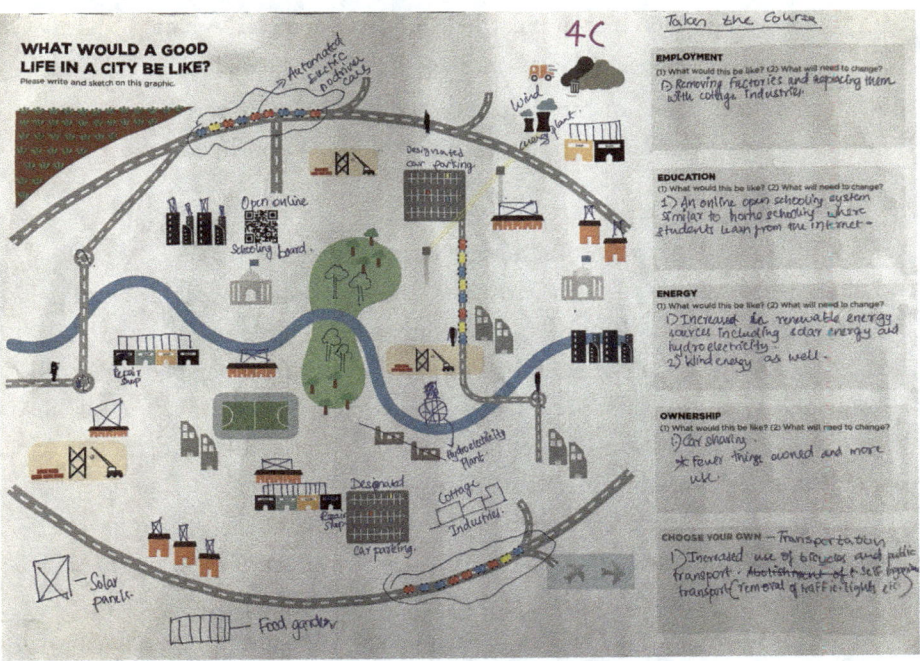

Figure 3. An example of secondary-school students' systems thinking
(Source: Heinrich & Kupers, 2019)

Across a number of evidence sources, including the activity illustrated here, the researchers did see shifts in all of the areas listed above. However, the path to this new way of perceiving the world was not smooth. Some students reported a level of discomfort with the learning when they were several lessons into the unit. Chapter 14 will come back to the complex dynamics of why we might feel discomfort when we are on the verge of significant new learning. These students were more comfortable by the end of the unit, once they had got their heads around the concepts, ways of thinking, and general academic usefulness of complex systems thinking. Reflecting on their learning after completing the unit one student is quoted as saying:

> I tend to think more of systems as complex rather than complicated, for instance, I currently see that although disciplines are taught separately and have different foundational approaches, they are all interconnected. You cannot do one without the other, which is especially evident for me, in the natural sciences. I think that I am becoming more aware of the fact that nothing in fact happens in a linear progression, but in flux wherein exists a complex web of stakeholders, cause and effects etc. (Heinrich & Kupers, 2019, p. 107)

Differences between complex and complicated systems are explained after the next example, along with the key idea of non-linear effects and events.

Making space for teaching about complex systems

I chose this example partly because the collaboration between a complexity scientist (Roland Kupers) and an educator (Sara Heinrich) seemed to make for a fruitful partnership. From their different perspectives and priorities, what did they see as necessary to include and why? I was able to discuss this question with Sara in several Zoom calls, and the book has benefitted from her insights and advice.

The knock-on effects listed above provide an early insight into the sorts of outcomes that would indicate that students can *transfer* what they have learnt to new contexts. They beg an important assessment question—what sorts of evidence might need to be gathered, and under what conditions, to demonstrate these effects? Chapter 12 addresses that challenge.

Above all, this example surfaces the tricky question of making space in an already-crowded curriculum. Heinrich and Kupers described four options.

1. Gradual introduction into each curriculum subject as these are reviewed.
2. Integration into resources that support teaching approaches.
3. Teach complex systems and systems thinking as a specific stand-alone module.
4. In the context of International Baccalaureate Organisation (IBO) schools, integrate complexity into the Theory of Knowledge course.

Each option has advantages and drawbacks. Heinrich and Kupers' preference was a stand-alone option because this allowed them to take an interdisciplinary approach. They were aware that some secondary teachers would not be comfortable doing this. (Indeed, working across different disciplines is a known challenge for curriculum integration and I address this issue in Chapter 10.) They also gave a second reason, arguing that students need to build familiarity with the concepts and language of systems thinking before it is introduced into traditional subjects, where it might be difficult to reconcile with traditional pedagogical practices. I'll come back to the pedagogical challenge in chapters 4 and 5. Meanwhile Chapter 3 outlines the "concepts and language" that need to be built. My hope is that, over time, all students will be gradually introduced to complexity across the years of schooling, so that this sequencing issue will not be as acute.

Working out how best to manage the COVID-19 pandemic

The final example illustrates how a model of a complex system might be used for an urgent practical purpose. The context of COVID-19 demonstrates how and why complex systems modelling can be a valued social and political capability. Figure 4 is a "causal-loop diagram" developed to support policy makers who have the responsibility to design effective ways to prevent or lessen the effects of the COVID-19 pandemic (Bradley et al., 2020).

Teaching for complex systems thinking

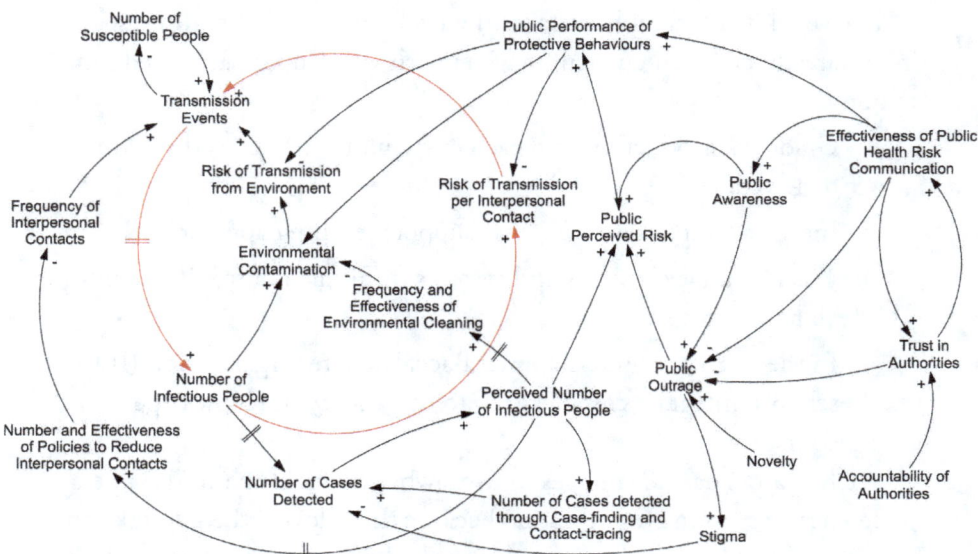

Figure 4. A causal-loop diagram to inform policy-making in response to the COVID-19 pandemic (Source: Bradley et al., 2020)

Notice the circular loop in red. This is an example of a "positive-feedback loop". Basically the effect being modelled will keep on getting bigger and bigger unless something happens to counter the positive feedback that is happening. In this example, the higher the number of transmission events, the more people will be infected, which in turn will increase the risk of them infecting others, which will cause more transmission events and so on. Left unchecked the result will be exponential growth in the number of cases. We have recently seen this dynamic play out in many nations around the world, with tragic consequences.

There's another example of a positive-feedback loop inside the red circle. As the number of infected people grows, so does the risk of environmental contamination, which then increases the number of contamination events. And so it goes. Notice that the word *positive* here does not mean *good*. There is an example in the next chapter where good positive feedback changes happen, but the COVID-19 examples of positive-feedback loops have very negative implications for human health because they cause exponential growth in the number of infections. By definition, exponential growth is not linear. As we've seen with COVID-19 infections in some nations, a gradual build up, if left unchecked, will suddenly explode into a runaway crisis.

When the positive-feedback loops have been modelled as exhaustively as knowledge of the virus allows, it is time to look for places to insert ways to break or counter the chains of activity just outlined. Specific actions that act to dampen-down activity include obvious measures such as introducing protective behaviours that stop the virus being spread by interpersonal contact. But they also include some less-obvious actions such as purposefully building trust in authorities (out on the right of the diagram) so that public outrage is lessened and people do not stigmatise those who are infected, making it more likely they will come forward to be detected and isolated from opportunities to spread the virus.

The researchers who developed this model are based in Northern Ireland and Kuwait. With our own experience of COVID-19 in New Zealand, you could trace a number of familiar loops that bring together seemingly unrelated things into one big messy, dynamic picture. Both science and social-science measures can be brought together within one model of a complex system. This is a real-world example of a complex system that has been playing out before our eyes. As the researchers note, many of the components of this model could be seen as "distant or disconnected" from one another until the "long chain of dynamic interactions" have been traced (Bradley et al., 2020, p. 1). I will circle back to the work of complex systems researchers in Chapter 15.

Two key insights to draw from this example

I chose this example to demonstrate why it is important that everyone understands the nature of complex systems dynamics. Looking at all the possible interconnections, it is not hard to see why the complexity modellers who crunch the numbers cannot give certain answers about what will happen next.

- When dealing with a novel infection, it takes some time before some factors such as transmission risks are clearly understood. A specific example here might be the ongoing debate about the risk posed by contamination of surfaces, particularly in relation to the import of frozen goods.
- Another source of uncertainty arises when modellers do not have ready access to robust data. Trust in authorities is an example—systematically gathered data about trust does not exist at the moment.

- A factor such as trust can fluctuate with the press of events and can be influenced by social media (which the model does not show).
- A small change in trust levels might have a bigger effect than anticipated if it leads some people to take risks or conceal infections.

When people do not understand the dynamic behaviour of complex systems, they can misinterpret the motivation of the modellers when the latter give a range of possibilities and say they cannot be certain. If this is seen as evidence of some sort of collusion, or deliberate obfuscation, people can all too easily fall down the so-called "rabbit hole" of conspiracy theories.

When I chose this example, I also planned to address an important conceptual distinction early in the book. I know Figure 4 looks a bit complicated, but *complicated* is not an appropriate adjective to use for this model. Think about a car engine, or a bicycle, say. In technical terms, both are examples of complicated systems. They have a fixed number of parts. Each part performs a specific function—take that part out and the system will be broken in some way. The connections between parts are linear and predictable. A car engine is surely more complicated than a bicycle, but both are called "complicated" systems because of these clockwork-like properties. By contrast the COVID-19 model shows a *complex* system because it has multiple interconnecting paths. Pathways between different places on the model are not linear and if one becomes blocked another could become stronger. The system can adapt and change over time. This is what is meant by complex systems being "dynamic". I will introduce many more examples as the following chapters unfold.

A quick note about scope

Working with complexity is relevant to every learning area of the curriculum. The examples above are set in the context of STEM subjects (Science, Technology, Engineering, Mathematics). Some also bring in the social sciences, working across subjects in an integrated way. Papers I read also included subjects such as geography whose systems span both physical and social sciences (e.g., Gilbert et al., 2019), the arts (e.g., Silva Pacheco, 2020) and language arts (e.g., Benson, 2020).

Some students taking part in these research studies were still in the lower primary school (e.g., Curwen et al., 2018). More of the studies were set in middle-school/secondary contexts and a substantial number were set at the tertiary level. The fourth example above is not set in a learning context per se. but in the world of work.

A pervasive theme of the studies is that developing *both* the knowledge and the dispositions needed to become a complex systems thinker is important for citizenship. Plate and Monroe (2014) used the term *systems literacy* as a focus for the assessment rubrics they developed (see Chapter 13). These rubrics describe what they would hope to see at "intermediate" levels of systems literacy, which they see as important for citizenship. Their "advanced" levels of systems literacy are needed for a wide range of current and future careers.

Expanding the scope of the definition, Benson (2020) introduces the term *systems citizens*. She defines this idea as follows:

> Systems citizens are *being* the changes they wish to create in the world, but they also know how to best pursue the systemic orchestrations required to bring those changes about. (Benson, 2020 p. 1, italics in original)

Brown (2019) uses the term *complexity competence* which is a "conceptual, experiential and practical form of knowledge" (p. 2). Again, there is an emphasis on *being* in complexity, but he also emphasises metacognitive awareness of complexity at work in the world. I'll come back to this idea in Chapter 12.

With this taste for complexity dynamics in place, I turn next to a more comprehensive discussion of the features of complex systems. The next chapter outlines the *knowledge* that students should have the chance to build across the years of schooling.

Chapter 2 reading guide

Complex systems are everywhere: systems inside systems, from the largest to the smallest of scales. The more researchers explore, the more complexities they find. Understanding what complex systems are and how they impact on our lives is important knowledge for all learners to gain over the course of their schooling. It is also important that students build dispositions to think in systems terms—doing this has both academic and citizenship benefits.

This chapter uses four different examples to provide a quick snapshot of complex systems and the challenges they pose for teaching and learning. In their different ways, they illustrate the lovely metaphor of an "intertwingly" world.

1. Cause-and-effect relationships in complex systems are often indirect and thus non-linear. Can you think of an example in your own life where an indirect relationship took you by surprise and made you think differently about causality? What sort of system was involved in this example?

2. An example from the Science Learning Hub[2] illustrates how scientists can also be taken by surprise when interactions are more complex than they initially thought. In what ways might learning about examples where experts are also taken by surprise contribute to students' growing "complexity competence" or "systems citizenship"?

3. The uncertain and unpredictable nature of outcomes in complex systems is another early theme to come through in this chapter. What specific challenges might these characteristics present for traditional teaching practices? Perhaps you could start by thinking about the example of the typical use of word problems in maths. Responding to the dilemma of "quick bright" students rushing to be first with the right answer, some teachers might say "we just need to use more wait time". In what ways might this response miss the point?

2   https://www.sciencelearn.org.nz/resources/75-decline-of-birds-and-pollination

4. The maths example also signals that teaching for complexity might help address persistent issues of equity and inclusion, by allowing some students the space they need to explore interconnections in a range of ways. Students who don't play "the game of school" by the same rules as others might have their chance to shine. What opportunities and risks can you see here?

Chapter 3
Features of complex systems

There is a body of knowledge about complex systems and how they work. The research literature is clear that this knowledge needs to be explicitly taught. By implication, all teachers needs to add this knowledge to their professional toolkits.

This chapter provides a broad overview of what is meant when we talk about complexity, or complex systems, or systems thinking, or any other similar term.

The first thing to note is that there is no overall consensus about how these things should be defined and what should be included. A systematic literature review of 75 empirical studies of complex systems in science education (Yoon et al., 2018) found many differences in how similar concepts are defined. The research team had to develop a coding system to be able to compare content of the various papers.

One Canadian research group has noted that this variation can be at least partly attributed to the purpose and context for a specific piece of research:

> Unfortunately, it is difficult to offer a concise definition of complexity because researchers tend to frame their meanings in terms of whatever they are researching. For example, synonyms for "complex systems" include "nonlinear dynamical systems" (mathematics), "dissipative

structures" (chemistry), "autopoietic systems" (biology), "healthy organisms" (medicine), "organized complex systems" (sociology), and simply "systems" (cybernetics). (Davis et al., 2015, p. 174)

This lack of consensus is not very helpful for teachers who just want to know what they need to know! In the detail that follows, I try to make practical sense of the overall picture by building up some key ideas that recurred in the papers that I have cited. As far as possible I avoid using unfamiliar technical terms such as some of those used by experts in the quote above.

Definitions that trace the development of the field

Some research teams explicitly differentiate between three different systems theories.

- **General systems theory** was developed first. It has a focus on the identifying structures and behaviours of systems. Relevant concepts include: components of the system; inputs and outputs; multiple levels of organisation; system boundaries and so on. (These features will be explained in more detail shortly.)
- **Cybernetics** came next. Developments in computing led to an expanded understanding of how systems can behave, drawing on network ideas and metaphors. Relevant concepts include: feedback; self-regulation; and equilibrium.
- **Dynamical systems theory** has a focus is on complex self-organising systems, with emergence, non-linearity, and far-from equilibrium states as added concepts. The complex ways in which systems interact could never have been figured out without advanced computing power (Weinberger, 2012). Knowledge of these dynamics is growing all the time as research centres and universities invest more resources into better understanding and modelling (Heinrich & Kupers, 2019).

All these ways of thinking about systems have something valid to contribute, but it is important to be clear about which type of theory is being used (Verhoeff et al., 2018). The second example in Chapter 2 illustrates the potential for confusion. Arguably the researchers were thinking about general systems theory when they developed their mathematical problem-solving examples (Salado et al., 2019). The systems they pictured in their cartoons had a number of variables that could interact in different ways, but they did not expect young mathematicians to be thinking about feedback loops and other types of system dynamics. In this specific context, the focus was on problem solving and the pedagogical argument made by the researchers was that systems thinking can support the development of problem-solving dispositions (see Chapter 9).

If, however, the focus is on *learning about* complex systems, or even understanding the dynamics of a specific complex system, then students do need to learn a number of concepts that might be new for them—and for many teachers. These concepts are outlined next, in my best approximation of general agreement about how they might be organised and summarised.

A summary of complex systems concepts

The concepts outlined here are broadly organised in the categories created by Yoon et al. (2018). Additional detail is provided from other papers, including one with a detailed conceptual framework developed to support more advanced studies in biology (Dauer & Dauer, 2016).

Structural features of systems

Every system has various **components** (sometimes called **variables** or **agents**). These can be identified and described. Descriptions also pay attention to structural features such as:

> **Connections:** Multiple interconnections between different agents are typical of complex systems. This is well illustrated in Figure 4 (p. 18).
>
> **Levels/scale:** Systems can typically be described at different levels of scale. The human body provides a familiar example. Underneath the surface there are multiple organ systems that constitute a level; look inside each organ system and there are different organs; look inside them and there are different tissues; and then there are individual cells and so on.
>
> A somewhat different example comes from a paper that explores challenges for learning about how particles behave in a phenomenon such as diffusion (Samon & Levy, 2020). At the macro level, changes caused

by diffusion might be visible (e.g., if a colour change is involved), or a new smell might become obvious. At the micro level, particles are moving around to create diffusion but they cannot be directly observed. As we will see multiple times throughout this book, this invisibility of features and behaviour at microscopic levels of scale creates challenges for learning about many different types of systems. As another example, the decomposition activities that keep materials circulating in ecosystems are invisible yet critical to ecosystem dynamics (Hmelo-Silver et al., 2017).

Lee et al. (2019) similarly note that many systems dynamics occur at scales too large or too small to be observable. That makes these dynamics hard for young students to grasp. They give an interesting example in the context of the water cycle as a system. Groundwater is distributed microscopically in soil, but both students and teachers are likely to think the term *groundwater* means an actual, visible body of water such as an underground lake.

Initial conditions: This idea describes how the system is organised at the start of a specific time sequence. Team games provide an interesting example. An effort is usually made to keep initial conditions in the game system as equal as possible: playing by the same rules; having the same number of team members in each side; having an approximately equal mix of abilities, and so forth (Storey & Butler, 2013). Initial conditions are important because they influence the ways the system might change over time.

Boundaries: It is important to define the boundaries or edges of the system being studied. This might seem obvious, but it is not always clear, and is a structural feature that is often ignored by education research teams (Verhoeff et al., 2018).

Inputs/outputs: Complex systems are typically open systems. Energy, materials, and objects can potentially enter or leave. These changes have an impact within the system, but the dynamics of the system at the time can also influence the rate of flow in and out. **Stock-and-flow diagrams** are used to model the dynamics of movement through systems. These will be explained and illustrated in Chapter 6.

Teaching for complex systems thinking

The short case study Open and Closed Systems (Box 1) illustrates a number of these structural features in the context of a very simple physical model of the water cycle. Throughout the paper I use many different types of models and stories as a response to the advice that it is not possible to fully capture the features of a complex system in any one representation:

> [In complex systems thinking] multiple schemas, analogies, models or case precedents are needed to capture and convey the meaning of a situation. (Dauer & Dauer, 2016, p. 3)

Box 1. Open and closed systems: Defining boundaries is important

The image below comes from an Assessment Resource Bank (ARB) item called *Where Did the Water Go?* Some years ago this sketch was repurposed from one of the very early ARB items and given a very simple "systems" focus.[3]

Students were told that the water in the open jar disappeared over the course of a week but the water in the closed jar was still present. After being asked to explain where the water went, the following question was asked:

Maria said "That one is just like the Earth's water cycle".

i. Whose jar was she talking about? (Tom's jar or Simon's jar)

ii. What do you think she meant? In what ways is it like the water cycle?

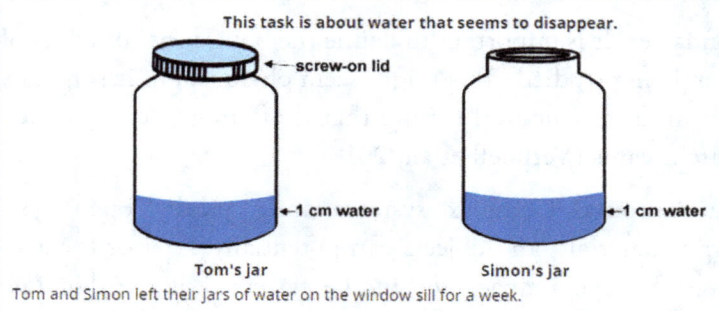

---

3   https://arbs.nzcer.org.nz https://arbs.nzcer.org.nz Note that a log-in is needed to access the ARB item bank, which means direct hyperlink won't work. This item is in the Science bank, and is called *Where did the water go?*

> At the scale of the whole earth, the water cycle is a closed system. There is a fixed amount of water, although it is constantly in circulation and in different states (water vapour, liquid water, ice). The trials of this item investigated how many students would choose Tom's jar and show they could think about the model in those terms by the explanation they gave.[4]
>
> This particular item was "very difficult" for Year 6 students and "moderately difficult" for Year 8–10 students. Looking back at this item, taking into account both scale and boundaries, I can now see that things are not quite so simple. At the whole-earth scale the water system is indeed closed. If we zoom into a specific ecosystem—and take care to define its borders—water will move through it at differential rates, in different ways, depending on multiple variables such as the weather, the presence and actions of animals, plant cover, and so on. At the ecosystem level of scale, the water cycle is open. Students who chose Simon's jar did not seem to know this however—they mostly said that evaporation only happened in the open jar and therefore this was most like the water cycle. The micro-scale dynamics of evaporation and condensation taking place in Tom's jar did not seem to be apparent to them.

The dynamics and processes that take place in systems

I turn now to the heart of the differences between general systems theory and dynamic complexity theories. These dynamics can seem unfamiliar, especially as we are surrounded by the familiar "clockwork" systems of machines and devices that behave in a predictable fashion, following a fixed sequence of events. As I outlined in Chapter 2, these are called complicated systems—they may in fact be quite simple, but the term is used to indicate that they are not complex. In Chapter 4 I will say more about the differences between complex and clockwork/complicated systems because many traditional pedagogical practices are based in clockwork-type assumptions. These assumptions are so familiar they are essentially invisible to us.

Interdependence/relationships: How the agents in a system interact with one another was the most commonly discussed feature of systems in the papers studied by Yoon et al. (2018). On one level this is the easiest

---

4   Each ARB item is, in effect, a mini research exercise. Students' responses are used to develop advice to support the use of the items in assessment for learning.

of the various systems dynamics to understand. Even young children can draw simple systems diagrams, using arrows to show connections. Food webs are a familiar example, although the direction of their arrows is counterintuitive because it traces the flow of energy through the system. On another level the dynamics of interactions can be hard to grasp because they happen *simultaneously*, all over the system, as opposed to the familiar sequential actions of clockwork systems (Dauer & Dauer, 2016). The dynamics of interactions can also *change over time* and how this change unfolds will at least partly depend on the *initial conditions* of the system.

Feedback cycles: Examples of positive-feedback loops were included in the COVID-19 example in Chapter 2. Basically, positive-feedback loops reinforce a change that is happening. Unless they are checked by countering interactions in the system, change could be exponential. Countering interactions come from negative feedback loops. They act to keep the system in a seemingly stable state (see the next group of concepts). Box 2 (below) provides a link to examples of feedback loops that are intended to be accessible to students from the top end of primary school. **Perturbations** are changes that the system cannot absorb in the existing feedback cycles. Box 3 outlines a famous example of a perturbation that had unexpected but very beneficial effects, specifically the reintroduction of wolves into Yellowstone National Park in America.

Box 2. An accessible account of feedback loops

The edition of the *School Journal* shown here includes a richly illustrated article about feedback loops, written by Matt Boucher, Deputy Principal at Thorndon Primary School in Wellington. New Zealand teachers should be able to readily access this resource. Supporting ideas for using it in class can be found on the TKI website.[5]

---

5 https://instructionalseries.tki.org.nz/Instructional-Series/School-Journal/School-Journal-Level-4-May-2020/Feedback

Box 3. The wolves of Yellowstone Park: A famous case of a major perturbation

In 1995 wolves were reintroduced to Yellowstone National Park. This is often cited as an example of a perturbation that led to ecosystem changes that took scientists by surprise. This perturbation actually restored dynamics that had been disrupted when the wolves were removed in the 1930s (another example of a perturbation). Wolves prey on elk, keeping their populations wary and on the move. In their absence, elk lingered along the edges of rivers in winter, causing considerable damage to willow and aspen trees growing there. Beaver populations dropped as the elk grazed and trampled the trees needed to build beaver lodges. Elk populations were again kept on the move after the wolves returned, and aspen and willow began to flourish. Over time beaver numbers increased and they built more lodges. Their chewing activities stimulated willows to grow more vigorously and the park became more verdant. Elk numbers actually increased too. This is an ongoing story and there is much more to it.[6] This snapshot is enough to illustrate the entangled nature of both **direct and indirect effects** that are typical of systems dynamics.

The Yellowstone example also illustrates **cascading effects**. One change triggers the next, and then that one starts more changes and so on. Cascading effects have been positive in the case of reintroducing wolves to Yellowstone Park, but they can also be negative. A series of cascading events can lead to a seemingly rapid collapse of the system in question. It can be said to have tipped over into a state of **chaos**. Chaos theory and complexity theory are closely related (Weinberger, 2019).

Self-organisation and emergence: These are particularly challenging concepts to grasp (Jacobson & Wilensky, 2006). They are not commonly included in research of systems education, at least in science (Yoon et al., 2018). The flocking of birds illustrates how these two ideas go together. A flock moves as one, with each individual bird keeping enough distance from its neighbours to avoid a collision. While on the wing, they appear to wheel and turn spontaneously. The behaviour we observe *emerges* from the small changes made by each individual as they respond to a simple set of rules. They *self-organise* to act as one group, even though there is no

---

6   Yellowstone Park provides its own account of this dynamic change here: https://www.yellowstonepark.com/things-to-do/wolf-reintroduction-changes-ecosystem

leader. Chapter 6 includes a simple physical simulation activity that gives students a direct experience of self-organisation and emergence.

Storey and Butler (2013) use a sports-related example to explain how emergent behaviour can result in the whole system "learning". Players respond to the conditions unfolding as the game plays out. If one player, or a small group of players, changes their behaviour to their team's advantage, the opposing player/team must adapt if they want to stay in the game. This is how skills increase and knowledge and use of strategies evolves. Eventually the patterns of game play that emerge will be different enough that the rules will also have to be updated. This dynamic also illustrates why complex systems are often described as **systems that learn**. The learning is in the ongoing shift of systems dynamics, as myriads of interactions play out, along with the feedback loops that respond to ongoing changes.

There is a risk that students who do not understand emergent processes will describe them in terms of direct processes, which are easier to see and explain (Dauer & Dauer, 2016). For example blood flow is a *direct process* with a specific cause (a beating heart). Diffusion is an *emergent process*, reliant on random movement of particles but the overall direction of movement may look similar to a direct process. Dauer and Dauer say that the most effective way to overcome this misconception is to specifically teach students about both direct and emergent processes, and the differences between them. Computer simulations can play a valuable role here, as we will see in Chapter 6.

Phase shifts: A system can undergo a phase shift which results in observable differences that become the "new normal"—until the next change. **Change-over-time graphs** are often used to plot these dynamics. Again, Storey and Butler (2013) use games to illustrate the concept. Games go through phases as playing patterns shift and adapt. Long periods of seeming stability can be followed by quite sudden shifts in response to seemingly small changes. One example they give is the sudden shift from offence to defence in many ball games. Another example is the change in energy levels displayed by the players at different stages of a longer physical game.

Systems memory effects: Just as systems can learn they can also remember. Memory is embedded in the structures and material conditions of *all* systems. This challenging idea is perhaps best grasped in the context of ecological memory. As one example, the trees that grow in an established forest are adapted to the conditions there. Their adaptations include ways to respond to disturbance cycles that have been overcome in the past. In areas where fires are common, the trees that prevail tend to have aerial seed dispersal: at least some seeds are likely to survive a fire and germinate rapidly afterwards. By contrast, forests that have no "memory" of fire don't have this type of adaptation. They are much more vulnerable to replacement by other ecosystems when they burn. This is what happened to swathes of New Zealand's indigenous forest cover when Māori, and then European settlers, fired the bush (Johnstone et al., 2016).

A related idea is that an unprecedented event which changes a complex system will impact on how the system responds *should such an event happen a second time in rapid succession*. This is a critical issue in the context of climate change, where ecosystems don't have time to recover from one perturbation before the next one is upon them. The second time around the system responds more rapidly and a non-linear pattern of accelerating change has begun. Hughes et al. (2019) provide an example in the context of coral bleaching caused by warming oceans. In their study, the coral managed to survive one season of warm seas and began to recover, but the very next year the sea was again too warm and the coral died. They conclude that the dynamics of ecological memory can lead to accelerating impacts of climate change.

States: How complex systems exist in the world
This third group of concepts provides a way of thinking about the "being" of complex systems in the world. Many of the ideas have already been introduced as various dynamics were outlined. Here everything comes together to make a whole—a whole that is *more than the sum of its parts*.

Diversity and order: In a clockwork system each part has specific unchanging relationships and role(s). By contrast, the agents in a complex system are more diverse, and so are their interconnections and their responses. Agents may act in different ways at different times. It is not possible to figure out how the system works, just by looking at the parts. Unpredictability is the order of the day.

Randomness: Many agents in complex systems behave randomly, but order emerges out the myriad random events taking place. In the physical sciences this is called **stochastic** behaviour. Again diffusion provides a good example. Particles move around randomly from an origin point. Eventually they will be spread randomly throughout the available space. The end result looks purposeful but it actually wasn't.

Far-from-equilibrium conditions: Complex systems can appear very stable for a long time. That's because of the constantly adjusting interactions that form an important part of the long-range memory of the system. This stability looks like a state of equilibrium. In fact some researchers do use that term, but many complexity scientists talk about systems being in a far-from-equilibrium state. They are always a bit off-kilter—the next perturbation could be the one that tips them into a phase change.

Resilience: The more diversity there is in a complex system, the more likely it is to maintain its far-from-equilibrium state, and hence the more resilient it will be (Zolli & Healy, 2012). The greater the diversity, the more opportunities there are for the system to learn and remember as it adapts to changing conditions. Resilience is closely related to long-term memory effects. A severe perturbation can destroy the resilience of a previously stable ecosystem. The example of forest fires, outlined above, illustrates this dynamic. In a severe wildfire, the seeds that form the "information legacy" of the system will be destroyed by the heat. The soil structure is also likely to break down, which changes the "material legacy" that the system provides for whatever grows next. The new system is likely to be less resilient and therefore even more vulnerable to repeat wildfire events (Johnstone et al., 2016).

Clearly resilience is an important concept for learning about sustainability. Yet Yoon et al. (2018) noted that resilience as a feature of complex systems was not mentioned in any of the science-education studies they included in their literature review.

The importance of non-linear change

The sorts of changes described in the examples in this section are mostly **non-linear**. They do not happen at a steady rate, or necessarily in one direction.

Exponential changes make a non-linear shape when they are plotted on a graph. The many different COVID-19 graphs we see in the media are a good example. The numbers of infections in a specific area will seem to build up slowly at first but then, quite suddenly, the numbers will explode and what seemed like a small problem will be out of control. You might have seen graphs of this type described as "hockey stick" graphs because of the shape they take.

Unpredictability is another characteristic of non-linear change. What actually happens when a complex system changes is contingent on how various possibilities actually play out. We've all heard about the part played by contingencies in the COVID-19 pandemic. Someone happens to interact with another person who is infected but doesn't know it yet. Before too long there is an outbreak to deal with. Those interactions can be personal, or totally coincidental—for example sitting on a bus next to someone who doesn't yet know they are infected. As another example, what happens after a person gets infected also has an element of unpredictability. We anticipate that "vulnerable" people will be more at risk, but sometimes, for no reason that is as yet understood, a younger, seemingly very healthy person will be badly impacted and might even die.

The maths word problem depicted in Figure 2 in Chapter 2 provides a less dramatic but very familiar example. How long it will take to get from the beach to the hotel depends on so many things that just happen in the moment: there could be a red traffic light; someone might cross the road without waiting for a break in the traffic; there could be road works; or a nose-to-tail car accident; and so on. The next chapter introduces the term *it depends thinking* as a strategy for introducing students to contingency and unpredictability as challenging aspect of complexity.

Why explicit teaching of this knowledge is needed

Verhoeff et al. (2018) discuss the "theoretical nature" of knowledge of complex systems. They point to a difference between empirical models and theoretical models.

- Empirical models, or concepts, name and describe things that can be observed and counted. Many of the *components* of complex systems are of this sort—for example numbers of beaver lodges in Yellowstone Park.

- Theoretical models constitute a different type of knowledge. For example, if we talk about a population "as a system" many of the interactions described above can be inferred from the observed behaviour—but they can't be directly seen or counted.

Verhoeff et al. say that the differences between these types of knowledge can create learning challenges for students. To overcome these challenges the teacher needs to build a bridge between these two different ways of modelling the world. That bridge needs to introduce an explicit language and way of thinking about complex systems that is accessible to students.

Creating a bridge between empirical and theoretical types of knowledge need not be made too difficult for younger learners. Verhoeff et al. draw on earlier research by a team that, over time, has created a simple systems learning framework called CMP (Hmelo-Silver et al., 2017):

- C = components—the various parts of the system
- M = mechanisms—how interactions inside the system work
- P = phenomena—what can be observed as a result of the system's workings.

These are broadly the same categories as the three I used to group systems concepts above. Students in classrooms that use the CMP framework learn to identify and name examples in all three categories. The order in which they do so may initially appear counterintuitive, especially if it is viewed through the lens of traditional teaching practices. More detail about using the CMP model is provided in Chapter 5.

What might "explicit teaching" look like?

My original intention was to write one chapter on pedagogy. As I began to sort and organise the research-based advice I found, it soon became apparent that two chapters would be needed. A very clear theme emerged concerning what won't work. I'm going to get that out of the way in Chapter 4, before turning to advice about what could work in Chapter 5.

Chapter 3 reading guide

There is a body of knowledge about complex systems and how they work. The research literature is clear that this knowledge needs to be explicitly taught. By implication, every teacher needs to add this knowledge to their professional toolkit.

1. In the Assessment Resource Bank (ARB) example, I reflect on an "aha" moment when I saw how important it can be to define the boundaries of a system you have in mind. Did you have any "aha" moments as you read this chapter? What new or expanded ideas will you take away from it?
2. Our beliefs about how the world "is" are largely tacit—we act on them without actually thinking about them. My hope is that this book challenges us to bring these types of beliefs to more conscious attention. What is your current understanding of the differences between clockwork/complicated systems and complex systems? On a continuum of understanding, where would you place your default (tacit) thinking right now?[7]

   Mostly assume a clockwork world   Mostly assume a complex world

3. Much of the research I found has been carried out by scientists and science educators. You can see that bias in this chapter. But complex systems are clearly also present—and important—in the social world. That's why I found the extended sports metaphor developed by Storey and Butler so useful. How helpful was the sports metaphor for your thinking about the dynamics of social systems? If we think about a class as a social system, and the teacher and the individual students as agents in that system, how might these ideas help with understanding the dynamics that make teaching so much more complex than many lay people seem to realise?

---

7   Even now, I think I would veer more to the left-hand side of the continuum in my *tacit* thinking, even though I don't want to. Traditional ways of thinking are hard habits to break!

Chapter 4

Traditional teaching practices can undermine complexity thinking

Familiar pedagogies from industrial-age schooling often work against the development of a disposition to understand the world in complexity terms. They are typically underpinned by linear models of cause-and-effect thinking, and may not make space for the sorts of critical/creative conversations that complexity thinking demands.

If teachers aspire to have students develop the habits of systems thinkers, they may need to make a number of changes to familiar pedagogical practices. This was a very clear theme in many of the papers I read. Drake et al. (2017) explicitly describe several key areas of teaching practice that are likely to need rethinking.

Making things simple reinforces linear thinking

The first practice identified by Drake et al. is what they call "reductionist" approaches, typically used to introduce students to new knowledge:

> Familiar teaching approaches typically try to reduce complex systems into their parts so they are easier to understand. Then we tend to look for linear cause-and-effect relationships between separate parts. Doing this is problematic because it ignores the essence of the dynamic whole that makes the system what it is. We need to find new ways

to keep the wholeness while still making the parts accessible. (Drake et al., 2017, p. 30)

A number of other papers also say that traditional approaches that break information down into easily grasped pieces are a significant impediment to developing complex systems thinking.

- Clark et al. (2017) say that "environmental education for children is still dominated by reductionist, information-oriented approaches." (p. 1). The implication here is that these approaches work against building the habits of systems thinkers, as well as making it more difficult to grasp the dynamic nature of systems as wholes.
- Dauer and Dauer (2016) argue that a balance needs to be found between introducing ideas that are too complex for students to grasp, and over-simplifying concepts in ways that lead to misconceptions about complexity. They call this tendency to over-simplify a "reductive bias" (p. 2).
- "Hyper-simplification" of complex concepts is a common approach in senior secondary physics textbooks. This form of reductionist, linear thinking can actually make the concepts of quantum physics harder for students to learn (Levrini & Fantini, 2013). I will come back to this argument when I explore the complexity of cognition in Chapter 14.

Reductionist ways of looking at the world are so deeply embedded in Western thought that they seem totally normal to those of us raised within that knowledge system, along with the idea that humans are distinct from and above nature (Sammel, 2020). Sammel suggests that deeply reflecting on our own ideologies is important intellectual work for teachers. This idea is expanded in several of the chapters that follow.

Rewarding quick right answers gets in the way of systems thinking

The second practice identified by Drake et al. (2017) is the expectation that students will show that they understand by giving "right answers to every question we pose" (p. 30). As we saw in Chapter 3, complex systems can behave unpredictably and multiple possible dynamics might need to be considered. Students need lots of practice in more contingent "it depends" thinking (see the example below).

Traditional expectations of quick, correct answers extend beyond teachers' question-asking practices. Chapter 2 introduced research in which Salado et al. (2019) critiqued the way that word problems are typically used in mathematics classes. When students are taught to quickly extract the numerical information from a word problem and convert the given numbers to a computational equation, they are being socialised to see the contextual information as clutter to be ignored. This research team called for a more open-ended exploratory approach to problem solving in which students work together to pose questions and problems, and not rush to the first answer:

> During problem-solving, systems thinking is characterized by challenging assumptions and searching for connections between elements of the problem and its context (Salado et al., 2019, p. 56).

Box 4. Introducing the idea of "it depends" thinking

NZCER's science ARB team coined the phrase "it depends" thinking when we found an atypical pattern in the responses of a small number of students from one school. Using the food web shown here, students were asked to choose three animals that would be impacted if a gardener used a spray that killed aphids and whitefly. They were asked to explain the nature of the impact in each case.

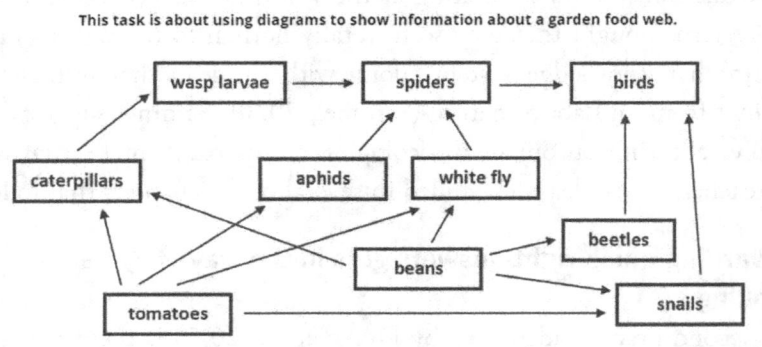

When the science team developed this task we also began from a position of traditional right-answer thinking, anticipating answers based on a linear reading of the various food chains in which aphids and whitefly are found. But then we found a small number of responses like this one:

> Spider – won't have the aphids or whiteflies to eat so will either die or prey heavily on wasps. Wasps – will either have no spiders eating them (because they died from lack of aphids/whiteflies) or will be eaten a lot more. (Year 10 student response to an ARB question).
>
> This Year 10 student clearly demonstrates contingent systems thinking. The team decided to call this "it depends" thinking. We went back to recode all the students' responses from all the schools that took part in the trial. We found several other examples and these were all from the same school as this student. It seems likely that these students had been explicitly taught to think in this more contingent way.

Linear cause-and-effect thinking is our habitual way of reasoning

The third practice identified by Drake and her colleagues is the neglect of explicit teaching of cause-and-effect thinking. Doing this sort of thinking requires:

> a disciplined blend of critical and creative thinking: critical because [students] need to look for hidden connections; creative because they need to look beyond the obvious to find non-linear links and interactions; and disciplined because this is not a case of anything goes. (Drake et al., 2017, p. 32)

They cite futures educator David Perkins (2014), who explains that our default tendency is to think in terms of simple linear causality and thus we don't think to teach for any other models of causality:

> Understanding complex causal systems is fundamental to navigating the contemporary world, yet complex causality gets no more than an occasional nod under the label of systems thinking. (Perkins, 2014, p. 211)

Perkins says that causal reasoning that involves multiple variables is largely ignored. This is a problem because thinking in terms of single, linear cause-and-effect phenomena is a habit most humans share and it is a habit we need to learn to break. Perkins suggests that teachers could begin to open up conversations by introducing instances where there is

- mutual causality—two things influence each other
- relational causality—a consequence depends on the relationship between multiple variables.

Deanna Kuhn (2020) addresses this same challenge from a different starting point. Like Perkins, she notes that adults prefer to explain phenomena in a way that is already familiar and acceptable to them. They tend to hold such views strongly and will discount explanations that do not fit with the current beliefs. She envisages an opportunity to help break this habit, and build different dispositions, by changing the way we introduce students to *conflicting accounts of events*. She says we need to find new ways to build and enrich students' cognitive representations of complex social issues, and proposes a research-based developmental sequence for doing this.

- Build foundations by exploring situations where several factors reinforce one another to contribute to an outcome (i.e., the factors are *additive*). One example might be the mutual benefits of healthy eating *and* regular exercise.
- Next, introduce *interactive* factors that have contrasting effects in different contexts. The example given is the variable impacts of potential carcinogens in different contexts. When exposed to a potential cancer-causing substance, some people seem to be more vulnerable than others. A person's genetic makeup is one factor, but so are other aspects of their lifestyle that might either mitigate or reinforce the risk.
- Once students can think in additive and interactive terms, they can be introduced to *divergent* causal explanations (where people hold genuinely different views of causality).

Kuhn notes that, when teaching adolescents about controversial issues, teachers tend to go straight to an exploration of divergent views. This is problematic because many adolescent students (and indeed many adults) hold what she calls "multiplist" views of the nature of knowledge—they see conflicting claims as "freely chosen personal positions of their holders, like pieces of clothing" (p. 28). Only once a person has reached an "evaluativist" stage of intellectual development, will diverging claims come to be seen as "a problem needing solving" (p. 28). In other words, if students who still hold multiplist views are introduced to contentious issues, they are likely to miss the point of the intended learning about *why* these conflicting views exist in the first place.

Kuhn's argument concerns our tacitly held views of the nature of knowledge. In philosophical terms these are called our *epistemological beliefs*. Using teaching practices that encourage *metacognition* is especially important in teaching for complex systems thinking because it helps surface tacitly held assumptions about knowledge and the way the world "is" (Sammel, 2020). Chapter 5 begins to address this challenge and it will continue to pop up in subsequent chapters. In particular, metacognition has considerable implications for assessment (Chapter 12).

Working on our own causal reasoning

If students hold tacit assumptions about the nature of knowledge and reality, it follows that the same is likely to be true for teachers who have not had the opportunity to deeply explore complexity thinking.

Table 1 (below) comes from research that explored the teaching of complex ethical issues to adolescent volunteers on a school-holiday programme (Yoon, 2008). The table was created as a coding scheme to use when analysing a range of data sources (interviews, conversations, models, written answers, and the like). Yoon was looking for a phase shift in how the participants understood the world—at what point did their view of how the world "is" switch from a clockwork view to a complexity view?[8] My purpose in reproducing the table here is to provide some signposts for teachers' own metacognitive exploration as they surface views that they are likely to have held tacitly.

Table 1. Beliefs about how the world is (slightly modified from Yoon, 2008)

Categories of component beliefs	Beliefs associated with clockwork mental model	Beliefs associated with complex systems mental model
Understanding phenomena	Reductive (step-wise sequences, isolated parts)	Whole is greater than the parts
Control	Centralised within the system	Decentralised (system interactions)
Causes/purposes	Single	Multiple

---

8 As an aside—this shift took different patterns for different students but most had made the switch by the end of the holiday programme.

Table 1 (*continued*)

Categories of component beliefs	Beliefs associated with clockwork mental model	Beliefs associated with complex systems mental model
Action effects	Small actions → small effects	Small action → big effect
Agent effects	Completely predictable	Random/ not completely predictable/ stochastic
Complex actions	From complex rules	From simple rules
How cause is attributed, or purposefulness of natural phenomena	Teleological	Non-teleological or stochastic
Ontology	Static structures, events	Equilibration processes

There are several technical terms here that need unpacking. The next two paragraphs also serve as a reminder of the concepts introduced in Chapter 3.

The second to last row of the table looks at how we attribute causes to events. If a person is thinking in *teleological* ways, the result of an event will be explained as something that was deliberately caused or inevitable. For example, Samon and Levy (2020) found that younger students will typically say that particles *want to move* during diffusion, and so the particles purposefully go to where there are fewer of them. But diffusion is an emergent and decentralised phenomenon—particles move randomly (stochastically) and as they spread out the pattern that we call diffusion emerges. Samon and Levy make a link between teleological thinking and anthropomorphism. This is the attribution of human motivation to non-human events. As they note, many children's stories are anthropomorphic, with talking animals etc. so it is hardly surprising that young learners think in these terms. Encouragingly, they argue that teleological thinking about invisible agents such as particles is a part-way step to an understanding of random/stochastic dynamics. The least developed explanations that children give are that something "just happens". Even if the reasoning is not correct in scientific terms, anthropomorphic accounts at least demonstrate the recognition that changes such as diffusion have causes that can be explained.

The word *ontology* comes from philosophy and is about our views of how the world "is". In a clockwork world, things are understood, and are expected to behave, in predictable mechanical terms. The word *static* is used in the table, but it doesn't mean stationary or action-less. Rather, it means "not dynamic". A complexity view of the world entails a very different set of ontological beliefs. There is an awareness that things may not happen in a linear, predictable fashion, and seemingly small changes can lead to very large events if they trigger a cascade effect, and so on.

Ontological beliefs tend to be held tacitly. We are typically not even aware that there could be a different way of thinking about how the world is. As we will see in Chapter 8, indigenous knowledge systems have quite different ontological foundations to Western knowledge. Assumptions that underpin indigenous views of the world vary from culture to culture, but they are in general closer to complexity ways of seeing than the Western knowledge system has been in the past (see for example Sammel, 2020). Changing these beliefs is hard work, which is why fostering metacognition is such an important part of teaching for complex systems thinking. The discomfort felt by Heinrich and Kuper's (2019) students in the earlier stages of the unit on complexity (see Chapter 2) is likely to be a symptom of a profound change in deeply help worldviews. Chapter 14 will circle back to what is happening in our brains when these types of profound shifts occur.

Is change even possible?

There is some empirical evidence that students can quite quickly expand their chains of informal causal reasoning when they are exposed to complex phenomena, using models that visually demonstrate the dynamics of those phenomena in familiar contexts (Petersen et al., 2018). Once teachers become aware of the limitations of traditional reductionist thinking, it also becomes possible to repurpose familiar pedagogies, using them in ways that can foster complexity thinking. How to do that is the focus of the next chapter.

Chapter 4 reading guide

Familiar pedagogies from industrial-age schooling often work against the development of a disposition to understand the world in complexity terms. They are typically underpinned by linear models of cause-and-effect thinking, and may not make space for the sorts of critical/creative conversations that complexity thinking demands.

1. This chapter has some rather confronting ideas about *typical thinking habits* that subconsciously underpin teaching practice. For example, it seems to be human nature to attribute a specific event to the single cause that most closely matches our current beliefs. This type of linear thinking is a hard habit to break because we are mostly not even aware of doing it. What other problematic thinking/teaching habits are suggested? Why might they be hard habits to break? What sorts of feelings do these ideas trigger for you?

2. The chapter begins to suggest pedagogies that could help both teachers and students break subconscious thinking habits. These include: looking for less-visible connections between events or things; practising "it depends" thinking where different outcomes are possible and plausible; and explicitly connecting parts into dynamic wholes. Which of these suggestions resonate for you? What might these practices look like in your classroom?

3. The chapter introduces the suggestion that the introduction of controversial issues should be carefully sequenced so that "multiplist" (anything goes) thinking is not accidentally reinforced. What other examples of additive causality (multiple inputs/same effect), and interactive causality (it depends effects) come to mind? Could you work with some of these examples in your classroom? How would you know if more secure foundations of causal reasoning are being built?

4. The importance of metacognition is another theme throughout this chapter. What metacognitive practices (if any) have you found to be effective? How could you build a bigger repertoire of effective metacognitive practices to tackle some of the thinking challenges raised? How could you support students to develop some of these practices?

Chapter 5
Pedagogies that support the development of complexity thinking

---

Once teachers are aware of the pitfalls to avoid, many traditional teaching strategies can be repurposed to support the development of complex systems thinking. Teachers need to pay specific attention to dynamic relationships between parts and wholes and use modelling processes to explore these relationships. Fostering metacognition is important, supported by the use of collaborative learning strategies.

Chapter 4 outlined the limitations of traditional teaching practices when the goal is to foster complex systems thinking. However, across the papers I reviewed there are also some clear indications of what can work. The various approaches and strategies described in this chapter will not be unfamiliar to many teachers. Along with having a sound personal understanding of complexity, what matters most is the teacher's *purpose* for using chosen strategies and how well the enactment matches that intent.

Keeping a dual focus on parts and wholes

Perhaps the most challenging requirement is to keep the learning focused on both parts and wholes at the same time (Staudt et al., 2018). As Chapter 4 outlined, traditional teaching practice typically breaks the whole down into parts, and may or may not bring these sequentially back to the whole once the parts are identified and described. Instead of these reductionist approaches, the intended learning needs to keep the big picture in mind if teachers hope to foster complex systems thinking. Even though doing so is challenging, it can be achieved with students as young as second grade (e.g., Curwen et al., 2018).

Creating various types of **visual models** of the system being studied is the most commonly suggested strategy for addressing the part/whole challenge:

- Models can be as simple as **pencil-and-paper drawings** that illustrate components of the system and annotate the interactions between them. Chapter 3 introduced a conceptual model called CMP (components; mechanisms; processes) used by one research group to guide students' thinking as they undertake such drawings (Hmelo-Silver et al., 2017). I will shortly provide a brief case study describing this research team's own use of this model.

- **Concept mapping** is a well-established critical-thinking strategy that can be used to "map" the components of a system. Boxes or circles are used to define the important components of the system, while arrows show relationships between these components. Annotations on the arrows explain the nature of the relationship (e.g., Huang et al., 2018).

Models on paper have the limitation that they can't show change over time, so they can only ever be a starting point for getting to grips with the dynamics of complex systems behaviour. More interactive modelling requires e-learning tools. These are explored more fully in the next chapter. For now, it is worth noting that computer simulations which support students to practice complex systems thinking are generally one of two types.

Top-down models (e.g., SageModeler) begin by introducing students to the overall dynamics of the system. They then use tools such as stock-and-flow diagrams (see Chapter 6) to drill down into the interactions.

Bottom-up models (e.g., NetLogo, StarLogo) begin by exploring the behaviour of individual agents in the system. They then simulate ways that agent behaviour, and relationships between different agents, leads to emergent behaviour in the system as a whole.

In one small pilot programme students were given the opportunity to work with both types of models (Staudt et al., 2018). Some students preferred one approach and some the other. However, whichever way students were introduced to the system, the developers identified a need to bring the two ways of modelling together and this proved to be difficult for them to achieve. At the time of writing their paper, the research team was beginning work on a third model that would integrate both top-down and bottom-up approaches. However they also hypothesised that students would be advantaged if they had experience of both types of models because they would be less likely to fixate on one type only.

Using modelling as an inquiry process

Helping students to create and continually revise visual models of a system can "transform the classroom into a space of observing, theorising, testing, discovering, and analyzing, thus linking academic learning to the real world" (Curwen et al., 2018, p. 3). This quote implies that **inquiry approaches** to learning fit well with systems thinking, with the proviso that both parts and wholes can be kept in the frame simultaneously.

Hmelo-Silver et al. (2017) say that "engagement with authentic science practices is critical to learning science and *may be especially important for learning about systems*" (p. 53, emphasis added). They note that modelling practices are at the heart of authentic science practices, and that scientists bring together a "top-down disciplinary perspective" (i.e., their theoretical thinking) and "bottom-up raw observations and data" (i.e., their empirical inquiries) (p. 53). The types of student modelling activities they describe use these same inquiry practices. As they build and revise their models students make their ideas visible and available for discussion in the class. The collaborative approaches to learning that they describe also serve to foster metacognition.

Box 5. Using the CMP model to support student inquiries

In a reductionist teaching sequence, the various components that make up a system would typically be introduced first. Once the "contents" of the system are familiar, the learning would move on to aspects of its behaviour. Hmelo-Silver et al. (2017) describe a teaching sequence using a simulated e-learning model of an aquarium that proceeds in the *reverse* of this typical order.

- First a phenomenon (P) is introduced—in their case study a fish has died in the aquarium. Why might that be?
- Debating possible reasons encourages students to think about mechanisms (M). These are not limited to this aquarium system—students think about why fish might die anywhere, proposing as many possibilities as they can think of.
- They then think about the sorts of evidence they would need to look for to test out these possibilities. These two steps with a focus on mechanisms help activate their prior science knowledge as a foundation for new learning.
- Now that students are aware of what sorts of evidence they might need to look for, they are finally in a position to build their model, including all the component parts they think it will need (C).
- Using e-simulation tools, the model has now become an inquiry tool that they can use to test out their hypotheses about why the fish died.
- Some potential explanations are ruled out as the evidence is gathered and debated. Students continue to refine their model "based on disciplinary knowledge and plausibility and parsimony to support/refute their ideas" (Hmelo-Silver et al., 2017, p. 58).

The way in which e-learning tools were integrated into this unit will be outlined in Chapter 6.

Focusing on processes/mechanisms, not just structures

A number of the papers reviewed I reviewed mention the neglect of explicit teaching of the *mechanisms* that drive the dynamic behaviour of complex systems. This neglect arguably reflects the more general neglect of explicit teaching of causal reasoning, as outlined in Chapter 4.

Three main arguments are made for specific teaching of mechanisms/processes.

- Learning about processes helps students to make connections between the different *levels* of a system. This is especially important when invisible dynamics at the microscopic level impact observed behaviours at the macroscopic level. Examples given include decomposition as the driver of nutrient cycling in ecosystems

(Hmelo-Silver et al., 2017) and diffusion, as already discussed in previous sections (Dauer & Dauer, 2016; Samon & Levy, 2020). Agent-based modelling tools were specifically created in response to this need (Jacobson & Wilenski, 2006).

- Explicit knowledge of mechanisms at work in one system is likely to be *transferrable* to other systems (Dauer & Dauer, 2016). For example, explicitly teaching the difference between direct processes and emergent processes in one system builds understandings that can be applied to any other system. This argument resonates with the call to explicitly teach theoretical concepts such as emergence (see Chapter 3).
- An explicit focus on mechanisms can be used to activate what students already know at the beginning of an inquiry process, as in the case study of the aquarium outlined above (Hmelo-Silver et al., 2017). Samon and Levey (2020) used the agent-based modelling tool NetLogo which allows students to explicitly investigate the dynamics of the processes taking place. In an inquiry process not dissimilar to the aquarium case study described above, students followed the behaviour of a single particle when other variables changed. Questioning was designed to bridge between the micro-level particle behaviour and the emergent macro-level pattern (i.e., the consequences of diffusion that can be directly observed).

Collaborative conversations and metacognitive processing

As chapters 3 and 4 outlined, theoretical ideas about complex systems and their behaviour challenge many of our tacit assumptions about how the world is. These tacit assumptions need to be made visible if they are to be challenged and changed.

Individual students can undertake metacognitive reflection. For example, Gilbert et al. (2019) say students need to pay explicit attention to definitions which are "the metacognitive resources required for the development of systems thinking skills" (p. 37). They outline a reflective process that follows a formative assessment activity in which students make a systems diagram and describe components and connections.

Once the model has been created students identify areas they feel most and least secure about. For their least secure aspects they pose five specific questions and identify types of knowledge that would help them improve their understanding/model. They also identify who might generate the type of knowledge they need to gain. They then write a one paragraph reflection on the insight they gained into the system they documented—and "how that insight was different to looking at the component parts or a single relationship" (p. 39).

Gilbert et al. worked with tertiary students. The school-level papers that raised the challenge of metacognition were more likely to emphasise *collective* reflection. The essence of the argument is that when a culture of thinking about thinking (i.e., metacognition) is established in the class, students may become more comfortable with debating ideas *as ideas*, and less concerned with being seen to be correct (Phillipson & Wegerif, 2020).

A wide range of visual tools has been developed over recent years to support systems thinking. Some of these tools have an in-built metacognitive component, and hence can provide a structure for shared conversations. Box 6 introduces one such tool—the *ladder of inference*.

> **Box 6. The ladder of inference: A visual tool to support metacognition**
>
> Our assumptions, values, and beliefs influence how we select data, interpret what is happening, and decide what to do. Our interpretations and decisions then feed back to reinforce (usually) our assumptions, values, and beliefs. We act on the basis of our interpretations, and our actions affect what data is available to us. So our ways of understanding and acting in the world create a self-reinforcing system, insulating us from alternative ways of understanding (The Systems Thinker, 2018).
>
> Figure 5 shows a version of the ladder of inference. This tool is often used in management education and many versions can be found online. I chose this version because it appealed to me visually and shows the thinking involved in moving both up and down the ladder.

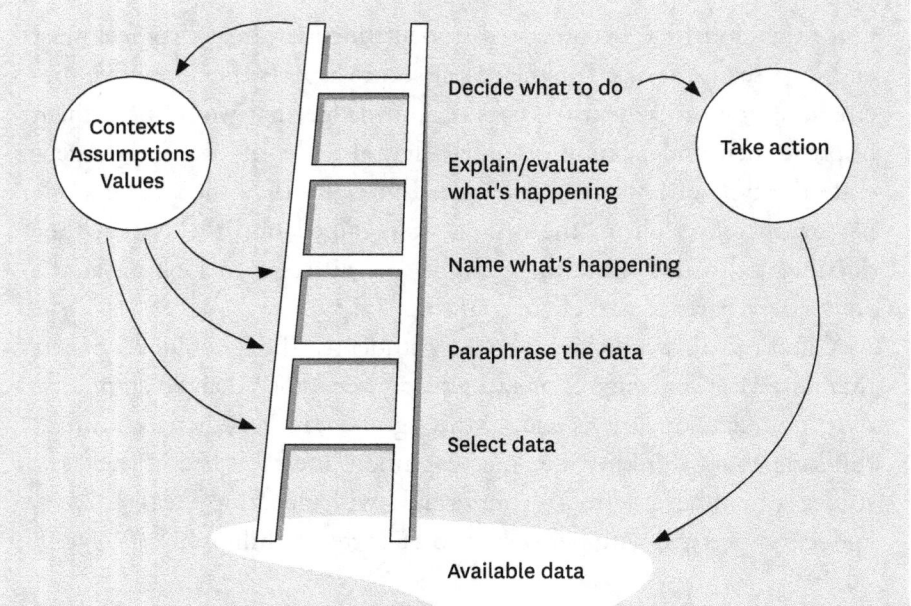

Figure 5. The ladder of inference reflection tool (After Argyris 1970; Senge, 1990)

This tool has a focus on how people work as metacognitive processors of information (Zohar & Barzilai, 2013). Teachers are perhaps most likely to encounter this as a means of reflecting on inferences they themselves make during their teaching. A short (4 minutes) video[9] produced in the New Zealand context (Robinson, n.d.) gives a clear introduction to this type of use. However, as we saw in Chapter 4, we also make many inferences about how the world is that might not serve us well in complex systems contexts. The tool can be readily adapted for this type of metacognitive conversation. For example, a range of social-studies ideas are included in a resource toolkit for IBO schools (International Baccalaureate Organization, 2020).

Collaborative conversations can make different ideas about complexity visible to all those who are taking part. Contexts where such discussions were described in the papers I read included school science learning (e.g., Hmelo-Silver et al., 2017; Yoon, 2008), environmental activism in citizen-science projects (Huang et al., 2018), geography education (Gilbert

---

9   https://www.educationalleaders.govt.nz/Problem-solving/Online-tools-and-resources/Ladder-of-inference

et al., 2019), visual arts education (Silva Pacheco, 2020), and a standalone module with an explicit focus on learning about complex systems (Heinrich & Kupers, 2019). No doubt such conversations are important right across the curriculum. They need to become part of a culture of thinking that is modelled by the teacher (Zohar & Barzilai, 2013). In sociocultural terms, we could say that the *affordances* of the context have a considerable impact on *opportunities* for students to build complexity thinking. One implication of these ideas is that the classroom environment, along with the learning opportunities on offer within it, makes an important contribution to the collective and individual success of learners. The importance of fostering the social/emotional components of systems learning will be discussed more fully in Chapter 7. Fostering a culture for building thinking dispositions is discussed in Chapter 9.

The situated *nature of shared metacognitive conversations*

DeVane et al. (2010) discuss how interactive computer games, situated in rich simulated contexts, provide opportunities for students to reflect on their meaning-making. Players take actions to meet their goals within the game, but at the same time the simulated system can support them to develop meta-understandings of the meaning-making model underlying that system. The process of "learning-thinking-doing" (p. 13) and the conversations students have with one another as they solve problems and make progress through the game, allow them to simultaneously try to understand the properties of the game system and the emergent effects they observe during game play. This description suggests that this type of complex game play can be simultaneously top down and bottom up, meeting the part/whole challenge outlined above. DeVane et al. call this type of activity *situated systems thinking*.

Yoon (2008) also introduces the concept of *collective metacognition* (p. 26). She describes strategies to make this an overt feature of the learning environment. For example, in one type of interaction, which the research team called a "cocktail party", students circulated to debate aspects of a complex issue while wearing an electronic display that showed everyone else their current position, which they were free to change as they talked. These interactions "required them to continuously monitor their own thoughts and understand their process of thinking to improve their skills" (p. 22). In other words their interactions stimulated metacognition. In a detailed analysis of the learning that occurred Yoon found that these shared conversations also served to amplify and accelerate students' access to new information—students said they learnt much more, and more quickly, than they would in a typical unit of work.

Leveraging complexity ideas to create new pedagogies

Chapter 3 introduced the idea that complex systems can learn and adapt. The "learning" is expressed as an ongoing shift of systems dynamics as the myriads of interactions within the system play out, and feedback loops respond to ongoing changes. The idea that learning itself should be treated as a complex system is discussed in Chapter 14. Meanwhile, this section concludes with ideas about using complex systems dynamics as a basis for designing effective learning interactions. Davis, Sumara and Luce-Kapler were early exponents of this idea (Davis et al., 2015, is the third iteration of their ideas on teaching as if the classroom is a complex system). The ideas introduced next draw on their work.

Storey and Butler (2013) argue that students need to be actively involved in making decisions about changes in game rules and learning conditions as they get better at a specific physical game/sport. Drawing on complexity concepts, they note that game structures and constraints evolve along with player abilities. Emergence leads to a need for new rules, and discussion about this can support players to be more aware of their own learning and growth. They also emphasise the need to help students gain awareness of their co-dependence on others during the game: they need to learn "how to adjust, adapt, invent, and play games that maximize opportunities for all" (p. 137) and that "learning only emerges in relation to others because it is situated within the system" (p. 139).

Storey and Butler say that teachers need to be aware that self-organisation occurs all the time within a game system. These dynamics are beyond the teacher's direct control, but they can and should influence play by "co-manipulating the constraints present in the game" (p. 137) and by giving feedback. They say:

> This shift from control orientation to recognition of learning as a biological adaptive process outside the teachers' direct control but within their influence represents a fundamental shift in emphasis from teaching to learning. (Storey & Butler, 2013, p. 138)

In a similar vein, Yoon (2008) points to similarities between complex system dynamics and classroom learning interactions. She designed her holiday learning programme to deliberately leverage several key principles of complex adaptive systems:

Leveraging variation: Systems are more resilient, and better able to adapt and learn, when they have sufficient variety. Seen in this way, the ideas and experiences that students bring to the classroom are a valuable source of variety. This might be especially important for learning about potentially controversial issues because "information enters the system in the form of opinions, beliefs, or potentially unexamined data by the learner" (Yoon, 2008, p. 8).

Harnessing interaction: Interactions between the parts of a system, in the form of feedback loops, are critical to the system adapting and learning. Yoon argues that students need to experience a series of exchanges over a given time, with the features of complex systems deliberately harnessed to support learning, in order for their thinking to grow and evolve. This is particularly important in the case of teaching and learning about ethical issues, where initial views are often implicitly but strongly held and it may feel threatening to have them challenged.

Some of the pedagogical strategies used by Yoon to meet these principles have already been introduced. They are listed below to re-emphasise the point made at the very start of this section—these are not necessarily new teaching strategies. It is the purposes for which they are used, and the ways that they are combined, that can foster complex systems thinking. Yoon's strategies for leveraging interactions included:

- face-to-face paired and small-group discussions
- participating in whole-group discussion in an online knowledge forum
- constructing risk-benefit charts examining tensions between environmental and social goals
- developing concept maps of relevant social, political, economic, and environmental stakeholders
- participating in whole-class events designed like the cocktail party conversations (see above)

- town-hall meeting simulations (with a special interest group at the heart of the debate)
- collaborating on the design and construction of a website for disseminating information.

Looking ahead

This chapter has raised the potential of e-learning tools to bring together parts/wholes and to model complex systems dynamics in a way that static learning materials cannot match. More detail about the design and use of e-resources is the topic of the next chapter.

Chapter 5 reading guide

Once teachers are aware of the pitfalls to avoid, many traditional teaching strategies can be repurposed to support the development of complex systems thinking. Teachers need to pay specific attention to dynamic relationships between parts and wholes and use modelling processes to explore these relationships. Fostering metacognition is important, supported by the use of collaborative learning strategies.

1. In what important ways do the key messages of this chapter reinforce and expand on those introduced in Chapter 4? For example, the importance of metacognition is one clear area of overlap with Chapter 4. What do you already do to foster a "culture of thinking" in your classroom? Could you tweak some of your favourite thinking practices so that aspects of complexity are a more visible focus of learning?

2. The chapter emphasises the role played by visual strategies when the focus is on part/whole thinking (e.g., creating models, concept mapping). What is your experience of using these types of strategies? What advice would you give to other teachers about using visual strategies, including pitfalls to avoid? Could you collectively create a resource bank of strategies to use in different contexts and for different reasoning purposes?

3. The use of systems models will be another ongoing theme throughout the book. Why might modelling by singled out as an effective pedagogy for fostering complex systems thinking? The chapter introduces the role of modelling in *student inquiry*, and directs attention to the importance of using the "authentic practices of a discipline". What does this term mean to you? What are disciplinary-inquiry practices? What makes them "authentic"? Where are these sorts of disciplinary-inquiry practices introduced in the overall curriculum? How might systems models be used as a starting point for disciplinary inquiries?

Teaching for complex systems thinking

4. Chapter 3 introduced the idea that a classroom is a complex system. Towards the end of this chapter, a number of strategies for harnessing this complexity are outlined. Which do you already use and how might you tweak them to foster students' complex systems thinking?

Chapter 6
Opportunities provided by e-learning resources

E-learning tools and resources complement paper-based and active simulations. E-resources can model systems dynamics in ways that make complexity concepts and thinking accessible for people of all ages. Different types of programmes model different aspects of systems. Ideally they will be used in combination. This is an active field of ongoing research and design work.

This chapter outlines how computer-based resources provide learning experiences that complement more traditional types of systems-thinking activities. Specifically, e-learning resources can model dynamic interactions in a system as these emerge over time. They also allow a system to be explored at multiple levels of scale. Neither of these things can be directly achieved with static media such as pencil and paper (Yoon et al., 2017).

There are a wide range of modelling programmes and resources that teachers could use, with more being developed all the time. It would be impossible to do justice to all of them, so this chapter gives a broad overview of the main types of learning opportunities that e-resources provide. As I decided what to include and what to leave out my thinking was influenced by several considerations. I have only included resources that are free for schools to use, and where I could find at least some research-based indications about how they might be used effectively.

Students need experience of both top-down and bottom-up models, and they need to see how these fit together (Staudt et al., 2018). *Top-down* models begin with a big-picture overview of the whole system then zoom in on detail of the interactions taking place. I address these first. Some computer programmes begin at the micro-level of a system. They show how the individual agents behave, zooming out to how this behaviour gets expressed by the system as a whole. These are *bottom-up* models and I discuss them second.

Introducing the "semi-quantitative" features of e-resources

Many of the resources intended for school learning come with built-in support features that allow students to do things that might otherwise be beyond their current mathematical capabilities. This is an important way to address a bit of a conundrum. Gilbert et al. (2019) note that the non-linear mathematics used to model systems dynamics is challenging even at the tertiary level. They say computation should not be introduced before students have a sound grasp of overall systems dynamics. But to get that grasp, they need to experience and play with systems dynamics! Good e-resources proactively address this conundrum. Where the mathematical processes are too advanced for school-age students, algorithms can be programmed in behind the scenes. Look out for these types of features:

- drop-down tabs give students a choice from a limited range of measurements for a specific variable
- slider bars allow students to set a quantity somewhere along the provided range
- arrows indicate comparative amounts (thicker if there is more of something/thinner if there is less)
- arrows can also show changes over time (if a line starts thin and gets thicker, more of the output is being generated over time)
- use of colour can indicate different types of impacts (e.g., green arrows for a change with positive consequences, red for negative)

- a drop-down menu provides students with choices of words they could use to define changes they see when they change the settings and run the model
- simple visual programming strategies allow students to use their computational thinking and build a simple a model that works.

Here's an example to give you a taste of how students might be scaffolded into writing a simple explanatory account when they run a top-down model of a system change. A screen shot in one paper (Damelin et al., 2017) shows the following display: "an increase in deforestation causes CO_2 in the air to *increase* by *a little*" (italics are the words provided in the drop-down display). The relationship is also displayed graphically on the same page of the e-environment, so students can see how the words and the graph pattern go together.

Top-down models

Top-down models begin with the big picture of the whole system, working downwards into the dynamics of interactions between the parts. For younger students, modelling a whole system might begin with a simple annotated drawing. This will introduce the components, but it won't necessarily reveal what the learners know about interactions between them, or help them to explore the mechanisms that drive those interactions.

One next step in a top-down approach might be concept mapping. Many teachers will be already familiar with concept mapping as a way of building up connections between interrelated ideas. Concepts (or components in the case of the CMP systems framework) are typically written in boxes. Arrows join the boxes as relevant and the nature of the connection is annotated beside each arrow. When students are first introduced to collaborative concept mapping, the process might be scaffolded by providing the concepts in boxes that can be physically moved around on a sheet of paper before being fixed into their agreed positions.

The causal-loop diagram of COVID-19 (Figure 4, Chapter 2) is another example of a top-down model. It is more elaborate than a simple concept map because there is a specific focus on feedback dynamics. But the model on paper is still a static version that cannot be tested in a simulation to see if it behaves as predicted. This is where e-learning resources enter the picture. I next look at a specific dynamic that can be modelled interactively.

Modelling stocks and flows

Stock-and-flow models put the emphasis on how things move around within the system—and perhaps into and out of it at the boundaries. Stocks are things that build up, or diminish, and flows represent processes that enable movement between stocks. A simple example is shown in Figure 6. In this example the stocks are places where water is stored and the grey arrows show the flow between the stocks.

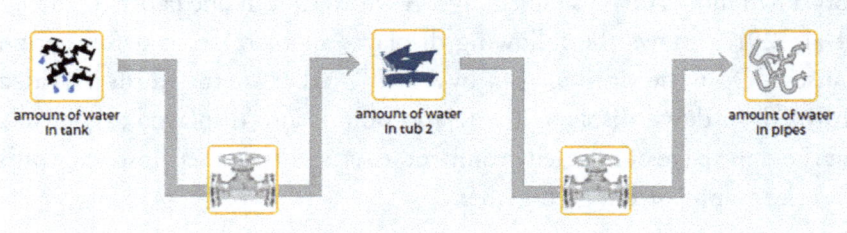

Figure 6. An example of a stock-and-flow simulation in SageModeler[10]

How much water will accumulate in tub 2? Even in this very simple example there are several sources of uncertainty—and hence opportunities to practise and model "it depends" thinking. The squares at the bottom of the model represent valves that could be open or shut. Even when they are open, they might not allow water to flow at the exact same rate. (Think about taps that are half-on vs. full-on.) Are the pipes the same diameter? If one was thinner, what impact would that have? The model lets students explore these relationships to test their predictions. As they run their models they can watch for change over time. Students might then apply what they have learnt to a real-world context where flows to and from a body of water have important consequences. One example might be lake with several rivers flowing in, changing rainfall patterns in the surrounding area, and a controlled outflow to a hydroelectric dam feed.

In the water example shown in Figure 6 the stock is an actual thing (water). But stocks can also be more nebulous things such as emotions.

---

10 SageModeler is one computer programme that does this. The Concord Consortium, which developed this software, has built a range of models for students to explore (Staudt et al., 2018). This and many other examples can be found at https://sagemodeler.concord.org/example-models/index.html

In the COVID-19 diagram (Figure 4) "public outrage" is a stock, and so is "trust in authorities". Both could go up or down depending on how policy, action, and events combine, with a range of possible impacts on the system as a whole.

Aronson and Angelakis (n.d.) say that many people struggle to understand the difference between stocks and flows when first learning how to convert systems diagrams to these models. They say a good way to test which is which is to imagine the whole system freezing at a moment in time. The stocks will remain unchanged but the flows will disappear because they are processes that can only manifest over time. Chapter 14 outlines how this thinking might be also used to help students learn difficult science concepts such as the transformation of energy from one type to another.

Box 7. An example of a physical simulation of stocks and flows

Stock-and-flow diagrams can be drawn on paper. They can also be modelled in simple interactive simulations that give students a more *embodied* experience of complexity. One case study with primary-school students comes from an international school in Switzerland (Boell & Senge, 2019). Students were learning about the complexities that have resulted in a net flow of refugees into Switzerland. They learnt about the pushes (conditions that drove them from their homelands and/or onwards into another place after coming to Switzerland) and pulls (conditions that drew them towards a specific place). They drew stock-and-flow diagrams to capture their understanding of these dynamics, but before they did that they played a simple dice game that simulated the complex conditions that drive refugee flows. The rules for the game are outlined here if you would like to try it yourself.[11]

Divide the class into small teams. Each team decides on a name and brief details for a fictitious country. They draw a shape on paper to represent its borders. They are then given 20 dice and a container big enough to hold them all for shaking/rolling. Each dice represents 100 people, so each team begins with a "stock" of 2000 migrants. You need additional dice to add to this initial stock, depending on how the game plays out.

---

11 These details were provided to me by Jacob Martin, a lead teacher of the International school of Zug and Luzern in Switzerland when this simulation was used, and now principal of an international school in Singapore.

Round 1: Experiencing the simulation

Each team rolls all the dice[12] and sorts them according to a set of rules that determine the meaning of each number

1. 100 refugees return to their home country. Remove 1 die
2. 100 refugees are deported because of a change in the law. Remove 1 die.
3. 100 illegal refugees are given up to the police by a people smuggler and deported. Remove 1 die.
4. 100 more refugees illegally enter your country. Add 1 die.
5. 100 more refugees are given asylum in your country. Add 1 die.
6. 100 family members of existing refugees are given asylum in your country. Add 1 die.

Once the team have sorted the numbers and calculated the overall impact on numbers of refugees in their country they record the result and roll again. After a number of rounds they draw change-over-time graphs to show how the numbers changed. If they get to 5000 additional migrants they must stop and have an emergency meeting to discuss what could be done to manage the situation in their country.

Round 2: The impact of conflict on refugee flows

In this round the first rule changes because conflict has broken out in a nearby country.

1. 200 refugees enter from a war-torn country. Add 2 dice

All the other rules remain the same but the impact of this one change will quickly become apparent. Once the second round is completed students can compare change-over-time graphs for the two rounds to see this impact visually. With this simulated experience of the dynamics of refugee flows, they should now be able to construct stock-and-flow diagrams that capture these dynamics in action in real countries.

---

12 Students were encouraged to roll the set of dice within the outline borders they had drawn for their country. This was intended to encourage them to feel a sense of connection to that place (Jacob Martin, personal communication)

Rethinking possibilities for an existing resource

For some years now, the science team at NZCER has harboured a desire to update a Water Cycle game-like resource we produced when *The New Zealand Curriculum* (Ministry of Education, 2007) (*NZC*) had just been released and the idea of key competencies was new for many teachers. The image below comes from a resource we produced at the time. The caption indicates that we were already thinking about the water cycle as complex but our main aim at the time was to disrupt traditional thinking about what the water cycle looks like, and to explore the limitations of typical representations. (We had the key competency of *using language, symbols and texts* in our sights at the time.)

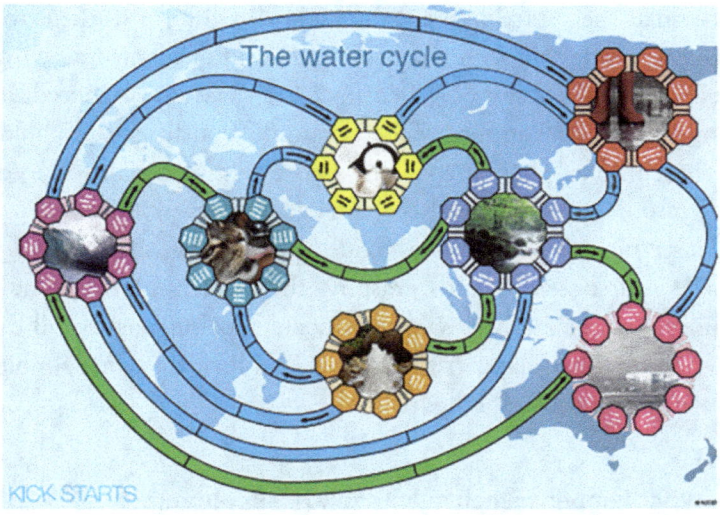

Figure 7. A game that could be repurposed as a systems-thinking resource

Part of my motivation for working on the literature review that underpins this book was to think about how we might update the game resource itself, with complex systems thinking in mind. The randomness of the movement of water particles was already allowed for: one way of playing the game is to stay on a circle for as many throws of a die as it takes to land on an exit point. Looking at it now, I can see that what we actually created was on the way to being a stock-and-flow model. Six of the circles are places where water collects (yellow=water vapour; magenta=sea water; turquoise=water in living things; pale orange=underground water; mid-blue=fresh water; pink=frozen water).

In systems terms they are stocks.[13] The red circle represents a mechanism in systems terms—rain, sleet, snow etc. are types of precipitation. We thought of this circle as a sort of driver at the time, needed to make the game work. The blue and green tracks are flows. We used the colour difference to depict one-way (blue) and reversible (green) flows. While this feature created useful learning conversations for students when they first encountered the resource, it would need to change now if we want to model stocks and flows, and especially if we aspire to build-in semi-quantitative features: each green track would need to be split in two.

Agent-based (bottom up) models

Bottom-up models take the behaviour of individual agents (components) in a system as their starting point. It can be difficult for students to grasp that complex *emergent* phenomena are the result of agents following simple sets of rules (Jacobson & Wilenski, 2006). I've already noted that the flocking of birds is a commonly cited example, but this pattern holds true for many complex phenomena. The two programmes introduced next provide resources to address this challenge. However you might want to consider giving your students a vivid experience of a *participatory simulation* as a first step towards grasping the concept of emergence. Box 8 outlines a simple activity where students become agents and directly experience what can happen when they all follow the same simple rule, without any one person being in charge.

> **Box 8. What happens when agents follow simple rules?**
>
> Here is an example of a participatory simulation that could be used to pave the way for agent-based computer modelling activities.[14] I first experienced this myself at an international workshop on complexity and was struck by how simple but powerful the experience can be. Each student becomes one

---

13 With the benefit of hindsight the images could cue some systems misconceptions about these stocks. For example most ground water is stored in microscopic quantities in rocks, not in open bodies of water (Lee et al., 2019).

14 This activity is included in the detailed plans of the unit trialed by Heinrich and Kupers (see Chapter 2). The full notes can be found here: https://www.rolandkupers.com/wp-content/uploads/2020/02/An-IB-complexity-module-for-the-Diploma-Programme-24.10.17.pdf

agent in a simulated system. As they follow very simple rules they have a concrete experience of emergence. A big outside space or hall is needed so everyone in the group can move freely around.

Ask each member of the class to choose two other students and to keep these choices secret. One is their "enemy" and the other is their "protector". Get them to move around so they always keep their protector in a straight line between them and their enemy. Everyone will mill about but very quickly they will find themselves in the perimeter of the space. Have them take note of this and then change the rule.

Explain that they now need to think of themselves as the protector. The two people they have chosen are enemies to each other. As the protector they need to stay in a straight line between the two people they have chosen. Almost immediately everyone will find themselves in the middle of the space. This simple rule change causes an emergent huddle, even though each person is just following the rule.

This simulation is a simple example that provides a fun and dramatic starting point for thinking about emergence. It does not allow for exploring interactions between different types of agents, or for change in agent behaviour over time. What it does do—and this is important—is to involve students in an *embodied* experience of complexity. They *sense* emergence as well as *think* about it. Chapters 7 and 8 will address the significance of adding sensing to thinking within the overall learning experience.

Agent-based computer simulations

NetLogo (Wilenski, 1999) is a free software resource that models agent-based behaviour in a system.[15] Just like the physical simulation sketched above, agents follow a small set of specified rules. As the programme runs, the behaviour shown by the system in response to the interactions of the agents becomes apparent.

A number of the papers I reviewed for this book had used or adapted NetLogo for the purposes of the learning they discuss. For example Hmelo-Silver et al. (2017) began their inquiry topic with a real aquarium in the classroom. In the first series of lessons middle-school students drew

---

15 The most recent version of NetLogo can be found here: https://ccl.northwestern.edu/netlogo/

pencil-and-paper models (top-down models) before they moved on to another series of lessons in which they used NetLogo simulations of the behaviour of various agents (algae, fish) in an actual pond ecosystem, as shown in Figure 8. Note that this learning was accompanied by specific teaching of relevant science concepts—photosynthesis, respiration, and decomposition.

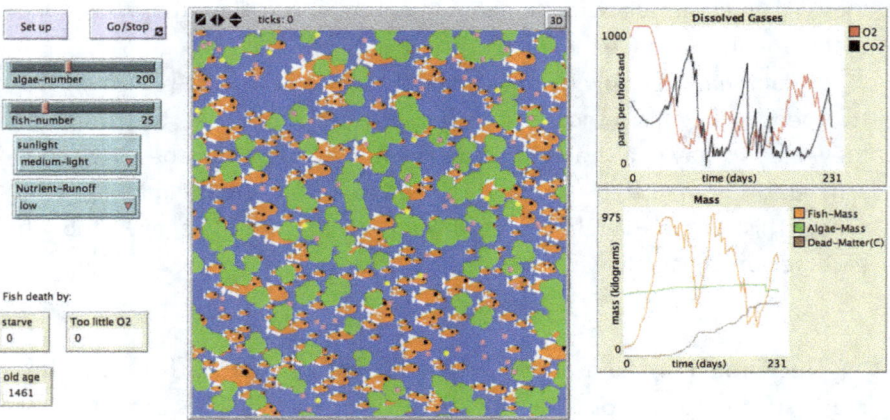

Figure 8. Netlogo simulation of an aquarium system (Source: Hmelo-Silver et al., 2017)

In this model system the fish are bright orange, the algae are green, and dead matter is depicted as brown blobs. As the simulation runs students can observe how the levels of oxygen and carbon dioxide change in the pond, and count the number of fish deaths when oxygen levels get too low. Biology teachers will recognise this as a model of eutrophication. If too many nutrients enter the pond, the algae reproduce so quickly that they overwhelm the capacity of the pond to provide the oxygen that they and all other living things in the pond need. Then they die (and so do all the fish etc.) and the activity of the decomposers even further reduces the amount of oxygen available. This is an example of a positive-feedback loop with very negative consequences (see Chapter 3).

Notice the semi-quantitative features in Figure 8 that allow students to experiment with changing the amount of several of the system components. Slider bars can change the number of algae or fish. Drop-down tabs are used to determine the amount of sunlight to which the system is exposed and the amount of nutrient run-off that enters the pond.

Other examples of agent-based modelling inquiries

In the fifth of seven sessions, Heinrich and Kupers (2019) used a ready-made NetLogo simulation called SugarScape (Li & Wilenski, 2009). This simulation models social inequality.[16] Like Hmelo-Silver and her colleagues, they used the simulation as one part of an extended learning sequence that made the modelling experience more meaningful when the time arrived to use the e-resource.[17]

Heinrich and Kupers began their 2-hour long lesson with a short reading that discussed the variety of purposes that can be served by modelling a system (Epstein, 2008). I will say more about this in the next chapter. They then showed a famous, very short YouTube video called *The Logic of Life*. This uses a carton of eggs to simulate agent-based personal choices that can unintentionally lead to social inequality.[18] The discussion that followed linked the simple simulation model in the video to the important warnings about the limitations of modelling, as set out in the Epstein paper. To illustrate, Epstein discusses the need to make assumptions while building the model, and the importance of being aware of the impact these assumptions will have on the outcomes that emerge when the model is run. Once this foundation had been carefully laid, the students were introduced to the SugarScape simulation. In our conversation about the unit, Sara Heinrich described this as the "most interesting" of the seven sessions.

Samon and Levy (2020) used a NetLogo model called GasLab (Wilenski, 1997) to explore students' understanding of diffusion. This simulation models the stochastic behaviour of gas particles in a closed system.[19] Note that there is an extensive library of NetLogo simulations that cross all learning areas of the curriculum. Many creative possibilities await teachers with the digital capabilities to use these resources.

---

16 There is a description here: https://ccl.northwestern.edu/netlogo/models/Sugarscape1ImmediateGrowback

17 As I was working on this chapter I was able to discuss the design of this lesson with Sara Heinrich during a very informative Zoom meeting.

18 https://www.youtube.com/watch?v=JjfihtGefxk

19 There is a description here: https://ccl.northwestern.edu/netlogo/models/GasLabTwoGas

Students building simple simulations

NetLogo can allow modelling of new systems and the most recent iteration of StarLogo (StarLogo TNG)[20] has some features that enable even quite young students to build and test simple system models of their own. Programming commands are presented as sets of interlocking blocks that enable students to build up command sentences without having to worry about getting the punctuation of the line of code exactly right (Klopfer et al., 2009; Yoon et al., 2017). Over a number of iterations the StarLogo software has also evolved to include a facility for 3D depiction of the system being simulated. Klopfer et al. (2009) say that this feature grounds students' modelling experiences in the world of games, and hence is more motivating for them.

StarLogo was developed at the Massachusetts Institute of Technology (MIT). Members of the MIT team have also developed an after-school programme called Project GUTS (Growing Up Thinking Scientifically) that uses StarLogo models as the basis for students' science inquiries. From these initial beginnings they have gone on to develop a series of modelling resources that fit a variety of curriculum topics. They also support a teacher professional learning group called Teachers with Guts.[21] The Concord Consortium[22] also uses StarLogo as its agent-based modelling tool.

Finding learning resources that will work for you

There is already an abundance of free programmes that can be used "as is" or adapted for specific learning challenges. Each has its strengths and limitations. Because this is clearly an evolving field of R&D activity, teachers might soon be spoilt for choice. In this fluid context, I thought it was worth reproducing a set of design principles recently published by one of the design leaders at the Concord Consortium. These are ideals that might be useful when making your own choices of resources, or adapting your own existing resources.

---

20 https://education.mit.edu/project/starlogo-tng/

21 https://teacherswithguts.org/resources/search

22 Concord Consortium is an American philanthropic organisation that provides free e-resources to schools. Many ready-built examples of systems models can be found on their website. https://concord.org/about/

Dorsey (2020) notes the limitations of resources that illustrate phenomena with single simulations. He says these are not sufficient if we want students to develop powerful insights into complex events and systems. His team at the Concord Consortium is now aiming to create virtual inquiry resources that provide "environments" in which students can explore complex phenomena as "active scientists" (p. 2). This aim resonates with the more general pedagogical principles set out in Chapter 5. He suggests that environments that support active student inquiry need to meet eight design principles.

1. Students can choose among multiple open-ended pathways (and learn from mistakes they make along the way).
2. Students are able to navigate feely among multiple interconnected levels of a system and identify key mechanisms that underlie phenomena they encounter.
3. The simulation environment reflects the natural world as authentically as possible and in ways that are appropriate for the grade level of the students.
4. Students can analyse and compare multiple, dynamic, linked representations of datasets they generate during the inquiry. These representations should make the underlying structure of the data clear.
5. It should be possible for students to collect artefacts (including data) that support the arguments they develop.
6. Students should have opportunities to return many times to the environment because multiple recurring sessions deepen knowledge and foster more sophisticated understandings.
7. The structure of the simulation should reflect the relevant structures and hierarchies of the natural world, from big picture down to an "under the hood" look at dynamics and interactions from which the phenomena emerge.
8. There should be low thresholds and high ceilings for potential activities within an environment so that learning stretch can be provided to the broadest possible range of students and to encourage increasing sophistication as students continue to explore the environment.

I imagine that aspects of these design principles will be familiar to many teachers from their more general professional experience. Some are more specific to complex systems thinking than others. Take principle 6 for example. Providing opportunities to deepen and enrich learning is important in any context, even if not always easy to achieve in practice. This is a generic argument for the principle. In the context of teaching for complex systems thinking there is another, more specific argument for the importance of this principle. If time constraints allow students to run a simulation only once or twice, they are likely to get similar results. They may well think that similar results would *always* be achieved—and hence that the future is knowable and predictable. In the same way that familiar and anticipated statistical trends can lull us into a false sense of complacency, having only a small number of experiences of complexity can deprive students of the surprise that goes with the unexpected. The greater the number of runs conducted, the higher the chance that an unexpected result will occur.[23]

Situating students inside rich learning experiences

In sociocultural terms, the *affordances* of e-learning tools are important. They allow students (and adult learners) to do things that would otherwise be more difficult for them (DeVane et al., 2010; Huang et al., 2018). Note that all the e-tools introduced in this chapter take either a top-down *or* a bottom-up approach to learning. Bringing both together can be done sequentially, but not simultaneously. As the next chapter outlines, a deep immersive experience inside a system might be an important way to bring students into a more holistic understanding of complex systems, their own place within them, and their rights and responsibilities towards all other living things and the planet as a whole.

---

23 My thanks to Jane Drake for this point (personal conversation).

Chapter 6 reading guide

E-learning tools and resources complement paper-based and active simulations. E-resources can model systems dynamics in ways that make complexity concepts and thinking accessible for people of all ages. Different types of programmes model different aspects of systems. Ideally they will be used in combination. This is an active field of ongoing research and design work.

1. This chapter has a focus on e-learning tools, but introduces them *in combination* with other types of learning experiences. How could the following *complement* one another to foster complex systems thinking? Can you think of specific examples that you currently use in your teaching, that you could tweak to make the notion of complexity more explicit?
    - Embodied simulations or game-like experiences.
    - Pencil-and-paper modelling tasks.
    - E-learning simulations.
    - E-learning inquiry environments.

2. What sorts of insights about the complexity of systems are conveyed by stock-and-flow models? Try creating a simple stock-and-flow model on paper. The chapter suggests a lake with several water inflows and outflows. You could also try playing the refugee game and then draw the flow of refugees into and out of a country as a stock-and-flow model.

3. Some people seem to think that e-learning tools can replace the teacher's role (at least in terms of the substantive learning experience). How do the examples in this chapter contradict that simplistic view? In what ways is the teacher's role critical to the success of the intended learning about complexity when using e-learning resources? It might be helpful to have specific examples in mind as you think about this.

Chapter 7
Situating ourselves *inside* systems

Students need to be supported to become aware of their own embeddedness inside multiple complex systems. Cognitive approaches to systems thinking are necessary but not sufficient. Our daily choices and actions are more intuitive than rational. Our responses to complexity are about our being in the world, not just our thinking. Students need support to build their dispositions to be and become complex systems thinkers.

In the introduction to this book I signalled that this chapter forms a sort of pivot point in the case I am making for the importance of supporting each and every student to become—and be—a complex systems thinker. I already knew it would be important to support students to see themselves as inside systems, not separate from them (e.g., Hipkins, 2019a). Yet every e-learning tool introduced in Chapter 6 could leave students exactly where they are most likely to already implicitly position themselves—outside systems looking in. This is not inevitable but I do think it is likely in the absence of deliberate attempts to disrupt traditional Western ways of thinking about the natural world. As an ex-science teacher, all the research accounts I had been reading seemed so normal to me that I didn't even question how I was structuring the content of the book. But I needed to address a couple of niggles that had begun to really hinder my thinking and writing.

My personal epiphany came as I struggled to think through the relationship between the e-learning tools introduced in Chapter 6 and those that are outlined later in this chapter. I was introduced to these next tools by international colleagues working on an initiative called "compassionate systems thinking". The migration story in Chapter 6 comes from a school working in that innovation network. I needed to reconcile this different approach with the literature already introduced. Once I saw that the primary intent of these next initiatives is to locate those who are using them inside webs of responsibility for their own choices and actions (as opposed to unpacking systems dynamics for their own sake) the disruption to my planned narrative trajectory became somewhat easier to resolve.

Introducing the idea of compassionate systems thinking

> In systems thinking, there is no "away", and this awareness of interconnectedness can underpin a life-long inquiry to understand ever more broadly the consequences of our actions, the root of all ethical behaviour. (Boell & Senge, 2019, p. 7)

Peter Senge is well known for his pioneering work on the use of systems thinking in business leadership contexts. A few years ago he worked with Daniel Goleman, who first described the concept of emotional intelligence, to address the challenges inherent in supporting students to build the *dispositions* needed to become system thinkers who care deeply about the health and welfare of others and our planet. They produced a small book called *The Triple Focus*, which identifies the need for systems thinking that links self, others, and the planet (i.e., a triple focus). They argue for an emphasis on the social and emotional aspects of learning in addition to the more familiar cognitive emphasis (Goleman & Senge, 2014). The *Compassionate Systems Framework* is a subsequent development (Boell & Senge, 2019). In the school context, this initiative seems mainly to have been taken up by the schools in the International Baccalaureate network (International Baccalaureate Organization, 2020), but is also spreading further afield, at least in the United States. The bullet points that follow are my paraphrasing of the argument for the approach taken in this initiative (Boell & Senge, 2019, p. 6).

- All of us are born with an "innate systems intelligence" that unfolds from early childhood, initially via our wider family relationships (whānau relationships for us in New Zealand).
- This systems intelligence remains "systematically underdeveloped" in mainstream education, given the traditional pedagogies that prevail (see Chapter 4).
- Exploring interconnectedness and change provides a powerful rationale for curriculum integration (see Chapter 10).
- Practical approaches and tools that could help teachers cultivate this innate intelligence are available but not yet widely used (see chapters 5 and 6, and later in this chapter).
- Both learners and teachers using these approaches and tools need to be actively engaged in reflecting on their ways of seeing and constructing their own models of reality. They also need the support of school leaders who understand what they are trying to achieve.
- Working rigorously with complexity and systems thinking has strong academic benefits more generally. The experimental unit described by Heinrich and Kupers (2019) is clear about this point, as is some other research not associated with the IBO network (e.g., Yoon, 2008).

You might be wondering why this group chose the term "compassionate" to characterise their work. They are explicit on this point, and on why they link it directly to systems thinking:

> we have come to talk of compassion as an essentially systemic property of mind: to cultivate compassion is to be able to appreciate the systemic forces that influence people's actions—"to walk in their shoes". Compassion goes beyond seeing a system from the outside—a kind of intellectual exercise—but actually feeling what it is like to being an actor within the system. (Boell & Senge, 2019, p. 5)

For me, this explanation resonated with conversations here in New Zealand about perspective-taking as an important facet of the *NZC* key competency of *relating to others*. But this group are clear that compassion is more than just perspective-taking, which they say can be a cognitive exercise that leaves the feelings untouched. They emphasise the importance of cultivating the disposition to *sense* the complexity involved in a system, not just to think about it, and they call this "sensing in". In the migration story (Chapter 6) students were given a powerful opportunity to sense-in when, towards the end of the unit, a small number of refugees were invited to come and tell their stories. For these mainly expat children, Switzerland was a wonderful place to be living and they had assumed that others coming to their city would want to be there too. It was very confronting for them to realise that their own reality was not that of these refugees. The homes of the refugees' hearts were no longer accessible to them, but were still where they would rather be, had circumstances beyond their control not made it impossible to stay.

I hope that this one simple refugee story will resonate with teachers who have successfully orchestrated powerful perspective-enlarging encounters for their own students. There is a lot of pedagogical skill involved in setting up a learning environment where it feels safe for students to be open to ideas and experiences that create dissonance for them, without being overwhelmed by emotions, or retreating too soon to safer certainties. The teacher must also model their own openness to learning. Boell and Senge use the idea of "social fields" to argue for the importance of creating safe relational spaces for these sorts of conversations because:

> the behavior exhibited by adults and perceived by students trumps espoused values and what is taught in the classroom, such as when adults teach students about respect but students do not experience being respected. (Boell & Senge, 2016, p. 4)

Boell and Senge note that there is a lot of wisdom and practical experience about how to teach for the sorts of encounters that prompt both new ideas and feelings. They say that learning needs to be: rational and intuitive; general and personal; conceptual and enactive/embodied; thinking and sensing (Boell & Senge, 2019, p. 6). This is a powerful set of both/and pairs, and I turn now to several more generic types of learning encounters might be employed to bring them together.

Pedagogies that bring thinking and sensing together

Describing pedagogies that meet this challenge could be a whole book in itself. I have chosen three examples that illustrate the richness of learning encounters that successfully bring thinking and sensing together, although those describing them do not necessarily use exactly those terms.

Mantle of the expert

A growing number of teachers in New Zealand are familiar with the pedagogy of Mantle of the Expert. In brief, well-established drama conventions are employed to give students a dynamic experience of a complex context. At times they look in on the problem or challenge as themselves. At other times, they are deeply immersed as actors in the system, taking up roles carefully planned for them by their teacher. Deep reflection is an in-built aspect of moving in and out of role, and teachers will often model this as they also take an active part in the learning that emerges. Many powerful examples can be found in a recently published book that also provides considerable practical advice on how to begin to work with this approach (Aitken, 2021).

Designing and making games

Games that are well designed and engaging typically work because features of complex systems are built in to the design. They have actors whose actions change specific variables (e.g., happiness, pollution levels) in different directions. There might be multiple levels of interconnections between different structural features and at different levels of scale. The occasional wild card might simulate the element of unpredictability in complex systems. They show emergence as the game play evolves, and so on. If this very brief summary interests you, a detailed analysis in the context of geography education can be found in Lux and Budke (2020).

Playing well designed and purposeful games is a very effective pedagogy for learning topics that come with in-built complexities. Playing such games is highly engaging for students, but building games for themselves is even more so (Bolstad & McDowall, 2019). Dr Bron Stuckey, who was a keynote speaker at the 2017 *Games for Learning* conference, argued that who does the designing of games matters:

> [Some] teachers get very excited about building the entire Roman Empire, or a village in London, and then having their students come in and play in it. Who did the biggest learning? The person who did

the design and built it. Because students, if they were doing it, are reifying their knowledge, their research, their understanding of what's happening there. (Bolstad & McDowall, 2019, p. 21)

The *Games for Learning* project explored several instances where students were able to design, prototype, test, and rework games for their peers to play. Teachers noted that initial failure is almost inevitable—even seemingly simple games are so complex that it almost impossible to anticipate everything that might happen when the playing action begins. They said that students seem to accept and work with failure more readily than they might do in other learning contexts. In this way, designing games provides useful opportunities to build dispositions to anticipate and tolerate uncertainty (see chapters 9 and 11). As the following comment from a group of Year 6 students shows, the designers also need to take the perspective of their prospective players if they hope to make a game that others understand and want to play:

[The trickiest bit was] making the rules, because everything had to balance out and it all had to go together … It got really complicated. We simplified the game a bit and thought about how everyone else would learn it, not just us. [How did you know how to do that?] Well we had play tested some other people's games, and only them who had originally made it knew how to play it, and it was extremely complicated to learn. In the end we thought, if it was really hard for us to learn, we should make our game easier for others. (Bolstad & McDowall, 2019, p. 28)

The purpose envisaged for a game can be adapted to any curriculum context. The final report of the *Games for Learning* project includes a number of examples if you are interested in following up this possibility (Bolstad & McDowall, 2019).

Bringing social-science methods into the sciences

Dana Zeidler and his colleagues are American science educators who have been researching the teaching of complex socioscientific issues for many years now. In a recently published paper, they strongly advocate for bringing social-science pedagogies into the science classroom when exploring socioscientific issues (Newton & Zeidler, 2020). In an interesting parallel to the compassionate systems work, this research group also identifies the critical importance of perspective-taking, with an emphasis on *ethical reasoning* alongside being able to understand other points of view (Kahn & Zeidler, 2019). They note that social perspective-taking challenges students to engage with "both the physical and sociocultural world where individuals must negotiate and coordinate between anthropocentric and ecocentric perspectives" (Newton & Zeidler, 2020, p. 15). In other words, complex issues must be considered from both ecological perspectives and human-centred ones.

The reintroduction of wolves in parts of North America provided the context for their most recent study. In Chapter 3 I also used this context to explain some important complexity concepts. I was curious to see how they moved beyond simply exploring the issue, as I did in Chapter 3, to immersing their student teachers right inside the complex context, prompting perspective-taking and ethical reasoning. They used a language-arts approach that involved reading and writing accounts of the impact the wolves might have from the perspectives of multiple parties: "livestock ranchers, environmental advocates, indigenous people, eco-tourism business people, hunter/sportsmen, and ecologists" (Newton & Zeidler, 2020, p. 7). Students then brought these perspectives to bear in a town-hall style meeting that debated how to get the best possible solution for both the ecological and the social-systems challenges they had been exploring.

Newton and Zeidler conclude their paper with a discussion of the importance of *social interactions*: student–student interactions in class; and student-invited stakeholder interactions (such as inviting refugees to talk to students). There are multiple ways these types of interactions can be achieved. Many teachers will already have favourite pedagogies they can use. It seems to me that one main challenge is keeping all the necessary pieces in play. There is a very clear message in Chapter 3 that there is a body of content knowledge about complexity that must be explicitly taught. That needs to be kept sight of as compassionate/perspective-taking pedagogies are also used to immerse learners inside the systems they are exploring, bringing the triple focus to bear (Goleman & Senge, 2014).

Some e-resources can immerse students inside systems

The very first example in Chapter 2 included the recommendation that students should learn to see themselves as embedded inside many systems if they are going to learn to take greater responsibility for their personal choices and actions (Clark et al., 2017). The team that designed the tool shown in Figure 1 commented specifically on the challenge that modern life poses for our awareness of, and connection with, the natural world (Petersen et al., 2018). They say that industrialised society:

> buffers us from directly experiencing many of the consequences of resource consumption that were more immediately evident to our ancestors. For example most of us in the developed world no longer have a direct connection with the production of food, the delivery of energy and water, and the processing of waste materials. Nor do we have a clear understanding of how our consumption choices affect important ecological systems on which we depend. (Petersen et al., 2018, p. 719)

Today's industrialised systems operate at a level of complexity and scale that creates "a complex array of local, regional and global feedback mechanisms" (Petersen et al., 2018, p. 719) such that the local affects the global, and vice versa. This dynamic multiplies the challenges of noticing the many consequences of our choices and actions. Petersen et al. say that our lack of action on climate change provides one illustration of the scale and complexity of this challenge.

The tool illustrated in Figure 1 (p. 11) was designed to monitor real-time electricity and water consumption at the community level, in order to get adults thinking deeply about their responsibilities to local ecological systems (Petersen et al., 2018). I find it encouraging that the wider team found a way to successfully adapt the tool for use with primary-school students (Clark et al., 2017). The two tools I introduce next were also designed to prompt adult awareness and deliberation. I have not found any equivalent research that explores their use with school students, but I am confident that creative, experienced teachers could find ways to do so.

Making choices about energy sources and consumption

Climate Interactive is an American non-profit think tank that originated as an initiative of scientists at MIT. They have created a suite of simulation tools designed to support community debate and decision making about energy choices and consumption. The tools are

> interactive, easy-to-use, and scientifically rigorous simulations that enable people to see connections, play out scenarios, and see what

works to address the biggest challenges we face. (https://www.climateinteractive.org/tools/)

Figure 9 below provides an overview of the climate-simulator tool. Notice the semi-quantitative features that come built in. Every variable can be altered using the slider bar, and the impact on carbon-dioxide levels will be immediately apparent. The intention is that groups can design and test scenarios for combining energy provision and consumption combinations that might be workable in their local context. What could make the biggest change in emissions given the way the variables play out at the moment? What might be changed to make the nation or community more sustainable?

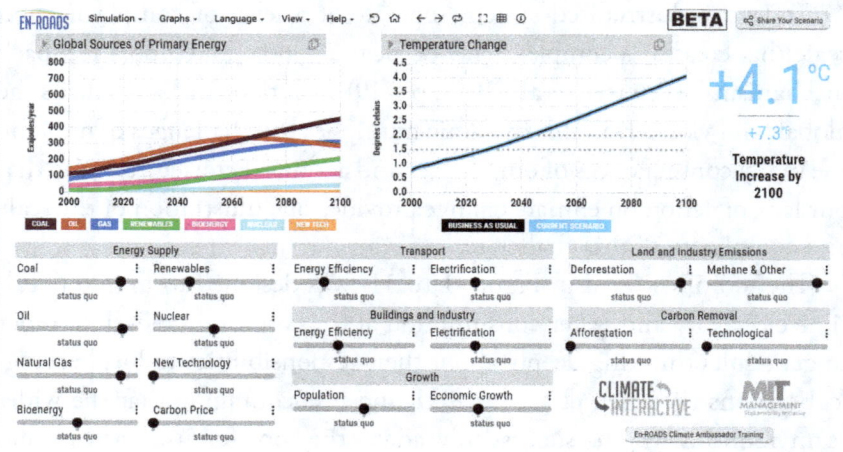

Figure 9. The EN-ROADS climate simulator
(https://en-roads.climateinteractive.org/scenario.html?v=2.7.19)

Notice that multiple interacting systems are built into the tool. There is no attempt to reduce complexity to its parts—all must be considered together. I would really encourage any teacher who is interested in using this tool to have a play with the simulator, and to check out the workshop materials that support it.

Bringing social and ecological systems together

A Good Life for All Within Planetary Boundaries[24] is underpinned by a major research project (O'Neill et al., 2018). This resource was designed

---

24 The home page can be accessed here: https://goodlife.leeds.ac.uk/

to allow adults to explore complex relationships between indicators of societal wellbeing and key markers of the biophysical health of the planet. The developers of the resource say that its purpose is to:

> foster a public discussion about the meaning of a "good life" and what it could look like in a world that lives within planetary boundaries. This discussion is vital – and urgent – because no country currently meets basic needs for its citizens at a globally sustainable level of resource use (https://goodlife.leeds.ac.uk/About/).

The impetus for the project came from the metaphor of doughnut economics (Raworth, 2017). A model of Raworth's doughnut is reproduced on the Good Life website. Meantime, picture this:

- Two concentric circles make the doughnut of the metaphor. The doughnut is coloured green because it represents the zone of sustainable life on planet Earth
- Inside the doughnut, a series of human needs are named in evenly spaced segments. These needs encompass: water; food; health; education; income and work; peace and justice; political voice; social equity; gender equality; housing; networks; and energy
- Around the outside of the doughnut, problems that arise when the earth's ecological ceiling is breached are depicted—again in evenly spaced segments. All of the following are included: ozone layer depletion; climate change; ocean acidification; chemical pollution; nitrogen and phosphorus loading; fresh water withdrawals; land conversion; biodiversity loss; and air pollution.

Notice that it does not deal with just one complex system, but with multiple systems simultaneously.

In this model the safe and just place for any society is the green band. Measures that don't reach the minimal *social foundation* show where inequalities are impacting unfairly on some people. Measures that exceed the *ecological ceiling* show where a society is living beyond sustainable limits. In this way, both the human and more-than-human elements of a nation's way of life are brought together and humans can never be outside the systems depicted. The interactive resources on the website could allow older students to explore their own country using different measures, and to compare it with other nations of their choosing.

I have looked at other websites and resources that also explore aspects of the circular economy. What struck me was that many of these resources picked one topic to explore—fashion say, or food.[25] They focus on the complexity of production and consumption systems within the chosen context. Obviously our actions and choices are implicated in those systems, and learning about them is worthwhile. But I found myself wondering how easy it would be to "other" myself—in essence to see the challenges as mainly someone else's responsibility, provided I made some effort towards more mindful consumption. Do they risk being "learning about" a system not "being inside" all the systems that sustain our life on the planet? I suspect this is a challenge we need to keep in mind as much as we possibly can.

A few thoughts on working with models

It should be apparent by now that building and working with models of different types is central to teaching and learning about complex systems. There are many different possibilities for working with modelling tools and strategies, including: games; physical simulations; drawings, diagrams, maps, graphs and other on-paper representations; and e-learning resources such as those introduced in this chapter and the one before. Given that the main focus of the book is on *teaching* for complex systems thinking, I once again circle back to pedagogy. Here are a few thoughts about using models in the classroom.

Models can prompt question asking

> Models can surprise us, make us curious, and lead to new questions. This is what I hate about exams. They only show that you can answer somebody else's question, when the most important thing is: Can you ask a new question? It's the new questions ... that produce huge advances, and models can help us discover them. (Epstein, 2008, paragraph 1.15)

Fostering curiosity and question asking can also help students to see themselves as being inside systems. Models that link complex systems to

---

25 The recently published resources of the Ellen MacArthur Foundation provide on attractive classroom-ready example: https://www.ellenmacarthurfoundation.org/explore

their personal lives are likely to be particularly effective here. As just one example, interactive models of predicted rises in sea level, a change predicted by climate modelling, have proven to be particularly effective in the compassionate systems thinking initiative.[26] Students want to know how places where they live might be impacted and this can lead them to asking many other questions, both about the processes that produce such models, and about the consequences should the predictions come to pass. New Zealand-based modelling tools of this sort have recently been produced—for example, a tool made by the Greater Wellington Regional Council.[27] Like many of the e-resources already introduced, this tool has semi-quantitative features that support question asking. Shifting slider bars leads to colour changes that become very obvious when zooming in on a specific area. These features allow students to explore different scenarios for the geographic areas with which they are most familiar.

Could you use the EN-ROADs or doughnut models introduced above to support question asking? How would you do that most effectively with your students?

Other ways to use models

In addition to question asking, Epstein (2008) suggests a range of other uses for models. One of these is *exploring trade-offs and efficiencies*. This is what both the EN-ROADs and doughnut models were designed to achieve when used in community conversations. "It depends" thinking could be a useful prompt here (see Chapter 4), when students use e-models to explore the potential consequences of different combinations of actions. Understanding *why* there can never be one "right" solution to complex problems is an important foundation for critical and nuanced action-taking. As I outlined in Chapter 4, linear cause-and-effect thinking is fostered and maintained by traditional teaching practices. Curious exploration of e-models can help students to break the habit of looking for quick right answers when there are actually none to be found.

Good models *make their assumptions explicit*. Epstein points out that we run mental models in our heads every time we think about

---

26 Jane Drake, personal conversation
27 https://mapping1.gw.govt.nz/GW/SLR/

the consequences of an action, but we seldom make our assumptions explicit when doing so, even to ourselves. By contrast, modelling activities can surface assumptions and allow them to be critiqued. There are some rich examples in Chapter 8, where scientists and indigenous groups with different starting assumptions work together. Asking students what has been taken as "given" in a specific model might be a useful way of unlocking curious and critical questions about the modelling process, not just the phenomenon being modelled. Surfacing and testing assumptions is one of a set of habits of systems thinkers developed by The Waters Foundation in the USA.[28] I'll come back to these habits in Chapter 9.

Models can be used to *decide what data to collect*. In many ways this is a logical corollary of question asking. While their modelling processes might be different, both scientists and social scientists use models to direct and focus ongoing inquiries. Older students could use models as a starting point for inquiries that lead them to seek data to answer their curious questions.

Looking ahead

As my own thinking and perspectives grew more complex, this chapter and the next chapter took shape together. I turn now to a discussion of complexity from the perspective of indigenous knowledge systems. I still regard the thinking I am about to lay out as personal work in progress, in the hope that being transparent about my own ongoing learning will be helpful to others on similar bicultural learning journeys.

---

28 https://waterscenterst.org/systems-thinking-tools-and-strategies/habits-of-a-systems-thinker/

Chapter 7 reading guide

Students need to be supported to become aware of their own embeddedness inside multiple complex systems. Cognitive approaches to systems thinking are necessary but not sufficient. Our daily choices and actions are more intuitive than rational. Our responses to complexity are about our being in the world, not just our thinking. Students need support to build their dispositions to be and become complex systems thinkers.

1. Near the start of the chapter Boell and Senge (2019) are quoted as saying that there is no such place as "away" and that developing an acute sense of our fundamental interconnectedness inside multiple systems is the root of ethical reasoning and behaviour. How many of our daily routines and activities are enabled by the presumption that there is an "away"? What are the ethical challenges of living like this?

2. Did thinking about the ethical challenges of living more sustainably create some dissonance for you?[29] What does the chapter say about the importance of fostering dissonance so that students' sense of complexity can grow? What are your own experiences of creating a safe space to challenge students like this? Which of the suggested pedagogies resonate best for you, in your teaching context? Are there others that you can think of?

3. In what important ways do the e-learning tools in this chapter differ from those introduced in Chapter 6? How might they complement one another?

4. This chapter provides the first real suggestion that some habits associated with Western thought systems will need to change if we hope to foster complex systems thinking. (Chapter 8 will say much more about this.) Specifically, we need to think less in

---

29 It did, and still does for me! I feel frustrated and powerless when I think about all the "away" mechanisms built into our daily lives.

either/or terms, and more in both/and terms. Near the start of the chapter there is a list of both/and pairs. Learning needs to be: rational *and* intuitive; general *and* personal; conceptual *and* embodied; thinking *and* sensing. What challenges for curriculum thinking, and for pedagogy, does this list raise for you?

Chapter 8
Learning from indigenous knowledge systems

Complexity is embedded in indigenous knowledge systems as a way of being and an overall worldview. It is intuitively *lived* via customary practices that have evolved over time, based on deep practical wisdom in local contexts. From indigenous perspectives individuals are *decentred* within the overall system, which does not privilege their interests over those of other living and non-living things, or the health of the planet as a whole.

As I was working on the earlier chapters of the book, I began to worry that I had not yet found any research that explored implications of learning about systems from indigenous perspectives. I could understand that to compare and contrast knowledge systems runs the risk of misunderstanding, or disrespecting, those that are seen as "other" to Western thought (see for example Sammel, 2020). I suspected that this was likely to be why accounts of complex systems thinking from indigenous perspectives have not been published in the places where NZCER's librarian and I had been looking for them.

One commentary on indigenous research methodologies points out that what it means to know something from a mātauranga Māori perspective will not fit neatly into Western knowledge structures, because indigenous perspectives are simultaneously a way of being and a way of knowing (Smith et al., 2016). This means that traditional research publications are not necessarily the chosen vehicle for sharing important new research insights. In the following comment, a Canadian research group add a related insight to this point:

> Many cultural concepts simply are not transferable to other cultures. Indigenous Sciences cannot be practiced within Western ontological assumptions and experiences. Even in this article we cannot avoid weighting our comparisons with Eurocentric meaning because we are writing it in English. (Hatcher et al., 2009, p. 145)

I too can only write in English, and my being in the world is strongly influenced by my education within the Western knowledge system and its associated beliefs. Nevertheless, with these limitations and risks in mind, I cautiously continued my search for insights about connections between indigenous knowledge systems and complex systems thinking from a Western perspective. Papers began to come to me from people in my networks who also have an interest in the questions I was asking (see the acknowledgements). I also began to find some of the sort of analysis I was looking for in literature published by scientists or public health workers rather than in education journals. From these scattered sources, I now endeavour to distil several messages that could be helpful for teachers with an interest in teaching for culturally responsive systems thinking. I find the themes that emerged encouraging. My hope is that they will be received in the same spirit, mindful of the caveats I have identified as part of my own learning journey.

Parallels between indigenous knowledge and complexity

> A systems perspective puts great emphasis on understanding the relationships between the components of a system, as it is the pattern of these relationships that determines the characteristics and properties of systems behaviour. It is in this focus on relationships and the meanings attributed to these relationships that we see the common ground linking Systems Thinking and indigenous Māori knowledge. They are two different bodies of knowledge, each with a long social and cultural history, but their commonalities, we believe, provide the opportunity to support and enrich each other. (Heke et al., 2019, p. 23)

So here is the first important message. There are strong parallels between indigenous knowledge systems and complex systems thinking. These parallels mark a shift in Western thought away from a dualistic view of humans as separate from and superior to all other living things (Hatcher et al., 2009; Heke et al., 2019; Sammel, 2020). This dualistic way of thinking is an intellectual inheritance from early Greek thought, and from the Enlightenment era when modern science as we know it emerged. This means that for many of us this human/other dualism is so deeply embedded as to be invisible unless it is explicitly challenged. Yet Western science also describes humans as co-evolving with nature, so there is a flaw in this deeply embedded logic:

> Therefore, as much as Science shows we are part of nature, it seems *normal* to perceive humans as distinct or above nature. (Sammel, 2020, p. 126, emphasis in the original)

Thinking from the perspective of complexity rejects this dualism. Chapter 7 addressed the urgent necessity to support students to perceive themselves as being *inside* complex systems, with an associated ethic of responsibility for our personal choices and actions. The key message was that there can be no "outside" perspective. This chapter reinforces the importance of challenging the common-sense view of ourselves as somehow above and outside nature, but then takes this argument a challenging step further.

Webs of relationships decentre individuals within the system

Here is the second theme that emerged from my personal quest for greater insight. I think it has huge implications for teaching and learning. The key message is deceptively simple: indigenous knowledge systems put webs of relationships at the centre of everything. As one example, the concept of whakapapa lies at the centre of Māori knowledge systems:

> Although whakapapa is generally defined as genealogy, it encompasses much more than that: whakapapa acts as a knowledge system that describes and contextualizes the origins and order of all things in the Māori world in relation to the individual. (Rayne et al., 2020, p. 519)

It is important to note that "all things" does not just mean all living things, which could be a meaning taken from another Western dualism (living/non-living). Whakapapa provides insights into "understanding the existence of and relationship between all animate and inanimate things" (Heke et al., 2019, p. 23). Furthermore, "species have whakapapa that connects them to their natural environment and to other species" (Rayne et al., 2020, p. 519). Everything is connected to everything and we are just one species within the networks.

The concept of mauri extends this critical difference to encompass the *spiritual* aspects of all things, again both living and non-living. Recent research of the educational implications of climate change recorded the following comment from a Māori sustainability educator:

> That's always been one of the drivers of kaitiakitanga, as best
> I understand it: are we being a good ancestor? And a conceptual
> description of kaitiakitanga is trying to revitalise mauri. And a
> definition I've heard of mauri is 'ma-uri' or 'for my descendants' which
> is a different framing, but the same word as being a good ancestor.
> (Bolstad, 2020, p. 8)

Notice how the concept of mauri is explicitly linked to a sense of felt obligation to care about and for webs of relationships in the natural world, as discussed in Chapter 7. A Samoan contributor to the climate-change project made a similar comment. As well as discussing mauri, this person also emphasised the importance of thinking beyond human timescales to consider our obligations in terms of whakapapa.

The decentring of humans within rich webs of relationships between all things is a feature common to indigenous knowledge systems. Other examples I found include the knowledge systems of Aboriginal Australians (Sammel, 2020) and of the Mi'kmaq people from the Canadian Atlantic coast (Hatcher et al., 2009). A number of important considerations when teaching for complex systems thinking emerge in association with this theme. I have chosen to outline three different considerations here. There will be more, but my aim is to at least make a start.

Locating students in their place

Place-based learning has always been important to indigenous people (Penetito, 2009). The individual is located within connected webs of meaning that quite literally put them in their place. Penetito says that culturally relevant learning experiences support Māori students to explore two key questions: *Who am I?* (a question of identity); and *What is this place*? (a question of ecological consciousness). Connections are "pervasive" in place-based learning and they include "multigenerational" and "multicultural" connections to community resources (Penetito, 2009, p. 7). Thus there is a need for learning to be explicitly anchored in the place where students live, and to take account of the deep local knowledge of that place held by indigenous occupants of it (Penetito, 2009).

Penetito says that place-based learning would be of real benefit to *all* students but two main types of impediments stand in the way. Like the many researchers cited in Chapter 4, he identifies invisible, reductionist, Eurocentric teaching practices as working against deep place-based learning experiences. He also notes that New Zealanders tend to treasure the ecology of wild and remote places, while ignoring environmental degradation much closer to home, and to which they contribute, whether they are aware of this or not. This second impediment echoes Chapter 7's dilemma of students being positioned outside systems looking in. The idea of "compassionate systems thinking" emphasises the importance of *sensing* connectedness, not just knowing about it.

What might sensing natural systems look and feel like in place-based learning? Wheaton et al. (2020) describe experiences that serendipitously push back against the predominant ecological focus on wild and remote places. They focus on "blue spaces", noting the importance of water in Māori cultural practice, and indeed to many New Zealanders as a source of leisure activities that enhance spiritual and physical wellbeing. I was struck by how the following quote from their paper brings together many of the arguments made by Penetito. Here a student teacher is talking about the experience of diving in the ocean:

> I would consider this experience both an adventure and a spiritual journey. I know my ancestors dived in these same locations long before me and I cherish that dearly. Through diving, I gain an appreciation and connection for the ocean and where I'm from ... My hapū (subtribe) in the far north is called Te Whanau Moana (people of the ocean, or family of the ocean). I feel a sense of connection to the ocean because of that and I feel I have a responsibility to my people to feel like I belong to the ocean. However the feeling isn't forced, when I am in the water, I feel grounded and entitled to say that I am a descendant of Te Whanau Moana. (Student teacher, quoted in Wheaton et al., 2020, p. 90)

The sense of deep immersion as an important condition for learning is also apparent in this quote, from a group of wellbeing researchers:

> an understanding of the ocean through the medium of surfing cannot be "learnt" as a theoretical approach—the ocean (or environmental representative) must be allowed to immerse you physically, spiritually and psychologically numerous times before an attempt at surfing can be made. (Heke et al., 2019, p. 24)

Both teams (Heke et al., 2019; Wheaton et al., 2020) discuss the Māori concept of atua matua as personifications and guardians of specific environments—they are the keepers of rich and subtle local knowledge that can only be learnt through deep immersion in the place they guard. One Māori sustainability educator who was interviewed for the climate-change research linked the idea of atua to being a more mindful consumer, as cued by doughnut economics (Chapter 7):

> [Our goal is to support] a holistic Māori view of environment, going back to our ancient belief system and atua, how everything is connected. And products, how everything that we use through the world comes from the natural world, and for us that comes from our atua. It's a way of helping people become more mindful and caring, and understanding why it's important to do all that you can to give back. (Bolstad, 2020, p. 32)

The same sense of rich, repeated learning encounters with a personally meaningful place can also be found in other indigenous cultures. For example, Hatcher et al. (2009) also point to the importance of providing rich experiences where students are immersed in webs of relationships at first hand:

> Indigenous ways of living in nature are strongly place based and the goal of Indigenous Sciences is to become open to the natural world with all of one's senses, body and spirit. (Hatcher et al., 2009, p. 143)

In their discussion of the knowledge system of the Mi'kmaq people, Hatcher et al. say that the basic premise of Indigenous Sciences is participating within nature's relationships, not necessarily deciphering how they work (which is the primary focus of Western science). There are a number of implications here. Not least, the types of experience described imply a need for frequent revisiting of places and experiences, as opposed to one-off visits. This resonates with the call for students to have opportunities to run systems models multiple times (see Chapter 6). The experience will be different each time, and each variation will be instructive of the complexity of systems interrelationships.

Hatcher et al. say that analysis of visual patterns can provide an important bridge between Western science and indigenous knowledge systems because it is a strength of both: "students need to be able to see the patterns in nature and recognize them as patterns" (Hatcher et al., 2009, p. 152). They introduce the term "pattern thought" to emphasise the intuitive nature of the meaning-making connections we build when closely and repeatedly observing/experiencing natural environments. They describe pattern thought as a form of complex reasoning, even if it is experienced as a hunch. "This way of knowing is relational, focusing on motion and change rather than things that are static" (ibid, p. 147).

As I see it, the sorts of rich experiences needed to build pattern thought have the potential to strengthen systems thinking dispositions. Box 9 introduces one very simple pedagogical idea, from the Canadian "ImaginEd" teacher education group:

Box 9. Introducing the idea of a walking curriculum[30]

The idea of a walking curriculum was developed by Gillian Judson with place-based learning in mind (Judson, 2018). The basic idea is very simple: students get out of the classroom and go for a walk of whatever time duration can be managed. They have a specific question or focus for their noticing, and they follow this up when they get back to the classroom. The website provides a range of suggestions, some more specific than others. Here's an example:

Shapes Walk—What geometric shapes (circles, squares, rectangles, triangles etc.) can you find outside? Collect & Organize: Do a tally of the shapes you find. How many of each kind? (ImaginEd website)[31]

This example creates a clear link to primary-school mathematics, both in the focus of the walk itself, and in the suggested follow-up activities. Other ideas relate to different curriculum areas and levels, so it is worth exploring them if you think this is something you would like to try. The website includes

---

30 http://www.educationthatinspires.ca/walking-curriculum-imaginative-ecological-learning-activities/

31 http://www.educationthatinspires.ca/2016/01/25/a-walking-curriculum-supporting-learning-through-focused-walking-k-12/

links to teachers' accounts of walks they have done with their students. In all cases suggested activities are designed to "deepen awareness of the particularities and meaning of place."[32]

Reframing problems using the idea of emergence

Heke et al. (2019) argue that decentring individuals in webs of interactions provides a constructive way of reframing thinking and learning about contemporary health issues. For example, rather than seeing obesity as a problem to be addressed, it is repositioned as an *emergent* phenomenon. In this way of thinking, obesity comes about when existing webs of relationships are not sustaining and healthy.

With this argument in mind, Heke and his colleagues used systems-thinking tools to construct a local curriculum in collaboration with one school community in the far north of New Zealand. Over several iterations, they designed a causal-loop diagram to show how a curriculum focused on mātauranga Māori knowledge could lead to the outcomes they wanted to see. The final version of this local curriculum model is shown as Figure 10 below.

The Health and Physical Education learning area of *NZC* already moves some way in this direction, with its focus on socioecological determinants of health, and health-promoting approaches to pedagogy. However in the atua matua model, rich webs of connections extend well beyond specific health-related factors to encompass all aspects of being deeply immersed in webs of natural relationships. Heke et al. suggest that this idea is more readily conveyed visually via a causal-loop diagram than explained verbally. Having this visual model allowed the group who developed the model to communicate what they were trying to achieve to others who needed to understand, for example the trustees on the school's board (Heke et al., 2019). The way in which the various different relationships come together to allow healthier lifestyles to emerge across the whole is made clear for all to see.

---

32 http://www.educationthatinspires.ca/walking-curriculum-imaginative-ecological-learning-activities/

Chapter 8: Learning from indigenous knowledge systems

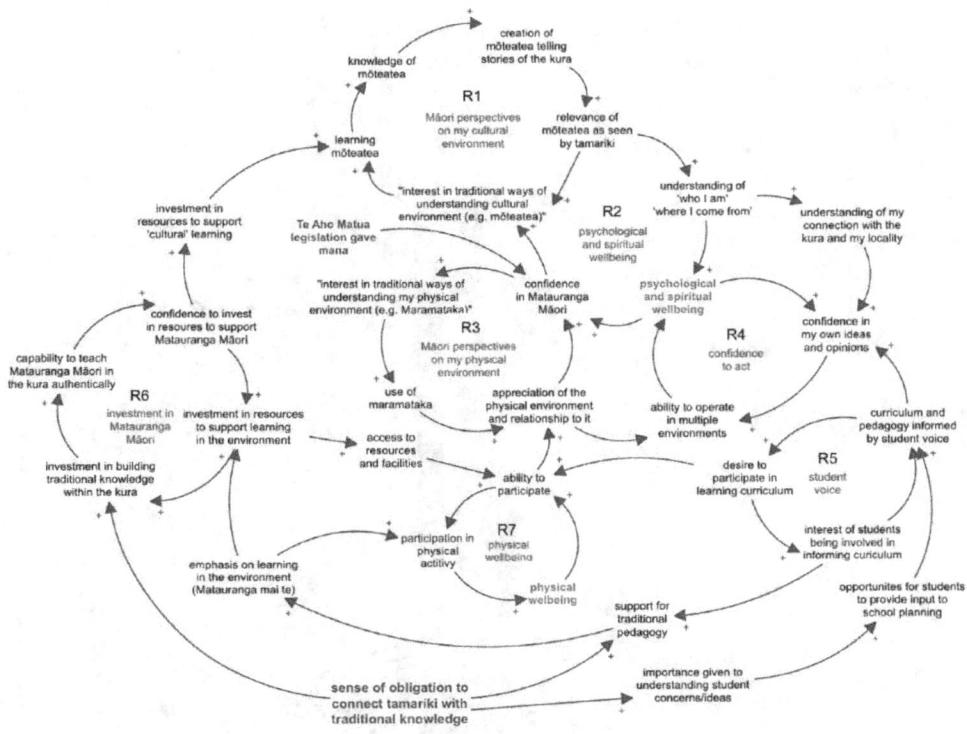

Figure 10. Using a causal-loop model to build a local curriculum centred on indigenous knowledge. (Source: Heke et al., 2019, p. 26)

Rethinking what is foundational and what is emergent

Sheridan McKinley, NZCER's Tumu Māori General Manager Māori of our Te Wāhanga Māori research group, drew my attention to an evolving conversation about what the doughnut model, introduced in Chapter 7, could look like from an indigenous perspective.[33] Figure 11 is the interpretation of the model developed by environmental scientist Teina Boasa-Dean to reflect the way her Tūhoe people would view the relationships between planetary constraints and social measures of the quality of life.

Notice that the spring of wellbeing which sustains all things, both living and non-living, is positioned resting on the oranga iho nui, the sphere of life and regenerative ecology. That sphere shields human social

---

33 https://www.projectmoonshot.city/post/an-indigenous-view-on-doughnut-economics-from-new-zealand

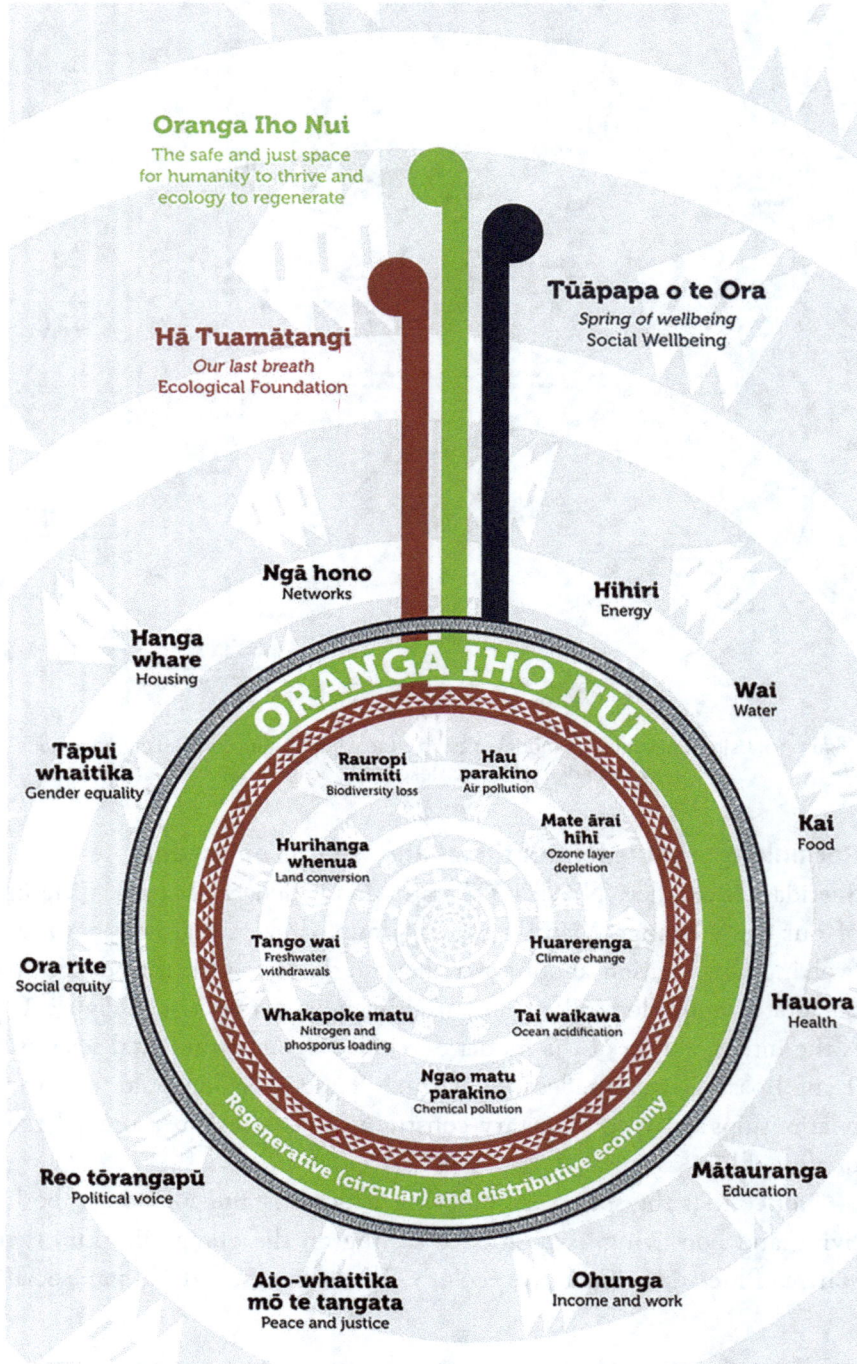

Figure 11. The doughnut model reimagined from an indigenous knowledge perspective (Source: Shareef, 2020)

wellbeing from unsustainable elements which spiral inwards, through the last breath, returning all towards a primal potential for life. However, that state of potential means that the sustainable life of oranga iho nui is no more.[34]

In her blog post, Shareef cautions that this version of the model does not represent the view of all tangata whenua (indigenous people of New Zealand). She is aware that there are other Māori perspectives on this same idea. Nevertheless:

> Whatever your worldview, it is powerful to recognise that two perspectives of the New Zealand context—Māori and Pākehā—can sit side by side. It is our hope that these two perspectives, and more, spark dialogue on our journey to build a better future. (Shareef, 2020)

This comment cues the third theme in the literature I found, namely the usefulness of systems-thinking tools for bringing different worldviews together in deep and respectful conversation.

The metaphor of two-eyed seeing

I turn now to the third theme in this chapter: it is important to find respectful ways to bring indigenous knowledge of complex systems and Western science knowledge together when addressing difficult challenges. Both ways of knowing have strengths. Decisions informed by both are more likely to anticipate and address the complexity inherent in the relevant challenge.

"Two-eyed seeing" refers to learning to see from one eye with the strengths of indigenous ways of knowing and from the other eye with the strengths of Western ways of knowing, and then using both of these eyes together. The concept was developed by the Mi'kmaq people from the Canadian Atlantic coast. Their long history of interactions with Europeans resulted in an awareness of the need for greater mutual respect for one another's ways of seeing (Hatcher et al., 2009). Hatcher and her colleagues have applied this metaphor to the design of place-based learning experiences, particularly for young Mi'kmaq adults attending their university, but with benefits for all science students.

---

34 My thanks to my editor, John Huria, for helping me to look deeper into this image.

How scientists use the metaphor to bring knowledge systems together

I first came across the metaphor of two-eyed seeing in a recent science paper written by a conservation team in New Zealand (Rayne et al., 2020). These scientists research and carry out translocations of endangered species. They say that Western science tends to take an either/or approach to what is foregrounded when making decisions about translocations (either ecosystem restoration or threatened species recovery) whereas indigenous-led approaches are more likely to incorporate both, taking a more nuanced approach. Customary practices, biocultural monitoring, and social mechanisms reflect a deep local systems knowledge that is highly sensitive and adaptable to novel changes, such as those being caused by climate change. Furthermore, mātauranga Māori knowledge and practices are likely to extend to species that are under-represented in Western science and conservation management (Rayne et al., 2020).

For the reasons just outlined, Rayne et al. say that indigenous knowledge systems should be centred when making translocation decisions. They use the concept of two-eyed seeing as a metaphor for bringing together relevant Western science knowledge, such as the use of genomic sequencing to ensure species diversity, and the deep ecological systems knowledge embedded in the cultural practices of tangata whenua. They also say that co-designed success indicators can catch a wider range of biocultural outcomes, and that customary practice is sustained or enhanced in the process.

Working actively with local iwi is implicated in the translocation work just outlined. This is also a strong theme in a collaboration between scientists employed by Landcare Research New Zealand Limited and local Māori (Montes de Oca Munguia et al., 2009). This Landcare group describes working together on challenges of modelling water quality in relation to patterns of land use. This is clearly a complex issue and the group use agent-based modelling as a tool to ensure that Māori cultural values are embedded in the models they built. They report that:

> Scientifically and culturally based indicators, along with community-based approaches, potentially provide an enriched understanding of the environment with each offering a slightly different worldview about the health of freshwater systems. (Montes de Oca Munguia et al., 2009, p. 2)

Montes de Oca Munguia and his colleagues stress the importance of being clear about the assumptions that are built into any agent-based model they create collaboratively. When a social-learning process is used to build a model, assumptions should become both *transparent and intuitive* to all participants. There is an interesting resonance here with the Mi'kmaq idea of intuitive pattern thinking.

> **Box 10. An accessible example**
>
> This report provides a high-level account of a partnership between NIWA and the Maniapoto Māori Trust Board.[35] A 4-minute video on the same webpage illustrates the interactions and mutual learning that took place between the NIWA scientists and members of the iwi.  They worked together to build a cultural framework for assessing the health of their freshwater. The term *two-eyed seeing* is not used, but this is an example of the type of interaction that could be used in the classroom.

A thought experiment in decentring ourselves

I now circle back to the importance of metacognition in learning for complex systems thinking (see Chapter 5) and of supporting students to work with models in ways that surface assumptions (Chapter 7). With that reminder of my main agenda in mind, I close this chapter with a short discussion of how these ideas might be incorporated in the practice of teachers on their own bicultural learning journey.

Alison Sammel is an Australian science educator who, like me, is on a personal bicultural learning journey (Sammel, 2020). In the following extended quotation, she undertakes a thought experiment about what it might look like to decentre ourselves when learning about the process of breathing:

> Students could explore how the air surrounds the planet and interacts with all aspects of the planet. They could explore how oxygen, a

---

35 The report and video can both be accessed here: https://niwa.co.nz/te-kuwaha/research-projects/maniapoto-cultural-assessment-framework

component of air, could be viewed as external to our body at one moment, but in the next, becomes a critical part of the very chemistry of our bodies. [This alternative approach] could show how this oxygen molecule can stay in our bodies for a short period of time or for the rest of our lives, and so becomes part of us, no longer classified as external or non-living. It invites students to reflect on some of the dichotomies that are positioned as normal: external versus internal, living versus non-living and so on, with the aim of opening up spaces for discussions that highlight similarities and connections rather than just perceiving differences. …. Students could explore how oxygen connects everything: how it is made by plants, used by plants and animals, and interrelates with all aspects of the planet. (Sammel, 2020, p. 140)

For me, this thought experiment created interesting resonances with the thinking that underpinned our design of the water cycle resource (see Figure 7). We designed one game-like use for the resource where students would throw a dice, going around and around the circle they were on until they landed on an exit square, at which point they could zoom off to a new destination. Revisiting this earlier work, I might now say that water was centred in our resource in the same way as Sammel centres oxygen in her thought experiment. This is not a huge step from traditional content teaching, but it does point to the challenge of undertaking a different sort of curriculum design thinking. Chapter 10 will address that challenge.

Another clear implication of both chapters 7 and 8 is that it is important to help students develop and strengthen their dispositions to think in terms of a world of complex systems in which we are ultimately decentred, notwithstanding the very evident impacts our actions have on everything else. It is easy enough to describe the fostering of dispositions a learning goal, but what might doing this actually mean in practice? That is the challenge I take up in Chapter 9.

Chapter 8 reading guide

Complexity is embedded in indigenous knowledge systems as a way of being and an overall worldview. It is intuitively *lived* via customary practices that have evolved over time, based on deep practical wisdom in local contexts. From indigenous perspectives individuals are *decentred* within the overall system, which does not privilege their interests over those of other living and non-living things, or the health of the planet as a whole.

1. This chapter identifies a number of thought patterns that are so embedded in Western knowledge systems that they can simply seem "normal" to those of us who live within a singular world view (me included). People in this position may not be aware that there are different ways of thinking about how the world "is" (i.e., our ontologies). Find and discuss examples scattered through the chapter, both in the text and the figures, which point to important differences between Western thought patterns and how indigenous people see the world. Do you find any of these examples more uncomfortable than others? Why might that be?

2. There are critical differences between mātauranga Māori and Western knowledge systems. When, why, and how could ideas embedded in mātauranga Māori, as well as differences between knowledge systems, be introduced to learners? If Western thought patterns are dominant in your own thinking, directly comparing traditional content with mātauranga Māori could risk appropriating indigenous knowledge into your more familiar body of knowledge. What could you do to avoid this dilemma if you have not had opportunities to be immersed in te ao Māori yourself?[36] (As I hope I have indicated, this is an important dilemma and learning journey for me personally.)

---

36 For readers from other countries te ao Māori broadly translates as "the world of Māori"—i.e., being immersed in the culture and way of being in the world.

3. The chapter identifies place-based learning as an important pedagogy for fostering intuitive awareness of the complexity of natural systems that sustain life on Earth. What reasons are given for singling out this particular type of pedagogy? In what ways does it encompass much more than simply *learning about* local places?

4. What if you began your place-based learning journey by deepening your understanding of your genealogy and your forebears' relationships to place? In what ways are you deeply connected to (a specific) place now? If you don't feel you have strong local relationships to people, places, and cyclic changes (e.g., seasonality) that surround you, how could you begin to build these? How could you contribute to fostering a sense of belonging and deep place-based relationships for your students?

Chapter 9
Fostering thinking dispositions by working on habits

Dispositions for complex systems thinking can be purposefully fostered by supporting students to develop relevant thinking habits. The relationship between dispositions and habits is itself complex. For teaching purposes, dispositions can be seen as "bundles of habits" which can be strengthened individually and collectively. Advice about how to do this broadly agrees with the pedagogical suggestions discussed in previous chapters. Practising in many different contexts is important and so is collective metacognitive reflection.

We need to educate *for* complexity if we want students to "make career choices and generally mature to become responsible citizens in an increasingly complex and interconnected world" (Heinrich & Kupers, 2019, p. 101). This phrase "educating for" implies that learning *about* complexity will not be enough. Students need to build the dispositions to be and become complex systems thinkers—to literally see complexity everywhere and make personal choices accordingly. Here are two more interviewees from the climate-change project introduced in the previous chapter:

> Once you understand climate change impacts in the global, connected sense, [you see it] everywhere you go, [and in everything] from the

supermarket to putting on the TV, to eating food, to drinking coffee is connected [to climate change].

I go ARGHHH because I see the bigger picture ... How do you get people to connect the dots? How do you [help them] connect colonisation to climate change? How do you [help them] connect mātauranga Māori to climate change? (Bolstad, 2020, p. 7)

I've included these comments to illustrate what a disposition to think in complexity terms might sound like. Such dispositions are easy enough to recognise when we see and hear them, but they are complex entities that bear further investigation. I begin the chapter with a brief overview of debates about what dispositions are, and in what circumstances they are amenable to change. Habits come into view as part of this discussion and I address the complex nature of their relationship with dispositions, drawing on several theoretical papers that helpfully survey previous literature and cutting-edge debates about this complexity.

Noticing and supporting thinking habits is one more in-the-moment challenge for teachers' daily work. I introduce a set of habits of systems thinkers that are designed to support classroom conversations, right across the curriculum. Many teachers will be aware of similar resources that support critical and creative thinking more generally. Perhaps the best known is Habits of Mind. I'll take a brief look at what is similar and what is different in these sets of habits.

The chapter concludes with a brief discussion of a suite of newly released New Zealand-based resources that are free for schools to use. The collection has been curated to support place-based learning, which Chapter 8 identified as an important pedagogical approach to culturally responsive teaching for complexity. In this chapter the lens of habit formation is added to the possibilities for using this resource.

A focus on dispositions and habits

Unravelling the relationship between dispositions and habits took me on an interesting side-journey. Looking back, I think I have taken this aspect of strengthening *NZC* key competencies too much for granted. What do we really mean when we say we want students to strengthen certain dispositions, and how do we think that will happen? Arguably, the closest I have come in the past is to draw on Guy Claxton's metaphor of helping students strengthen their "learning muscles" (Claxton, 2018). Just as physical muscles can grow stronger with exercise and practice, so can mental muscles. I still think this is a helpful metaphor, but it is a generic one. In the specific context of complex systems thinking, what actually needs to be strengthened and how? With chapters 4 and 8 in mind, it probably won't surprise you to hear that there are some tacit features of Western thought to unpick as we address this deceptively simple question. The Chapter 8 reading guide has already explored habits of Western thinking that can be hard to break because they are largely invisible to those of us who did not have access to a different knowledge system with which to make comparisons. What else might just seem "normal" unless it is closely scrutinised?

One important point is that dispositions are *not* fixed personality traits, though we often act as if they are. Writing in the context of initial teacher education, Nelsen (2015) says that such a view would be highly problematic because all prospective teachers would be selected on the basis of dispositions they already displayed. (The assessment dilemmas raised by this point are discussed in Chapter 12.) Extrapolating from this point, a fixed-dispositions argument leaves us with nowhere to go if we agree that being able to think in complexity terms is important for life beyond school. If we do think this, then *all* students should be entitled to have opportunities to learn and practice complex systems thinking.

Nelsen addresses the argument about fixed vs. developing dispositions by going back to the writing of John Dewey. He says that Dewey sometimes used dispositions and habits interchangeably (i.e., as if they were synonyms) but in other papers he invoked subtle differences. Drawing on the cues Dewey left in his prolific body of work, Nelsen suggests we should think about dispositions as "bundles of habits". This provides a practical way for teachers to think about how to influence dispositions. Teachers can work on smaller habits, building these as intelligent habits which are responsive to context, rather than fixed habits that are not able to be adapted. With this recommendation in mind, let's take a quick look at the nature of habits, through the lens of recent complexity research.

The complex nature of habits
The view of habits as fixed and unchangeable is also widespread in Western thought—they get a bad rap! For example, Nelsen notes that "we

often refer to habits pejoratively as we describe actions we wish we did not repeatedly enact" (2015, pp. 87–88). Writing in the context of addiction studies, Ramfrez-Vizcaya and Froese (2019) say that this negative view of habits stems from the mind–body binary that is deeply embedded in Western thought. They say that thinking about habits in this way is associated with behaviourist theories of learning that were largely discredited in the 1950s when cognitive-science research began to emerge. Now, neuroscience and social psychology studies have changed the picture again. Since the 1990s habits have become an interdisciplinary research focus across multiple fields, including psychology, neurosciences, philosophy, political science, organisational studies, marketing, behavioural economics, and transport studies. Ramfrez-Vizcaya and Froese discuss findings from these multiple research fields to build a nuanced and non-binary view of habits. Like other complexity researchers, they use the "4Es" to summarise contemporary theories about habits. The aspects described by each E interact to make a responsive, emergent whole:

Embodied: Habitual body memory allows past experience to be re-enacted implicitly but also to be tempered by context. Bodily memory of habits provides enough stability that new experiences can be encountered with the sense that you know what to do, even if you have not done exactly that thing before.

Embedded: There is a complex relationship between embodied memory and conscious awareness and control. One example might be using personal verbal hints and maxims in a conscious effort to get the body ready for action in a high-performance sport context.

Extended: Memory integrates social and material aspects of the context into cognitive responses. In this way, the context is not just a trigger for habitual responses. Instead it should be seen as an integral part of a distributed cognitive system. As Guy Claxton would put it, learning involves the mind–body plus books, devices, other people, memory and imagination, and so on (Claxton, 2008, 2015).

Enactive: Habits can be understood as mutually dependent: they comprise a self-sustaining adaptive network. Individual habits interact in complex ways to generate dynamic patterns of behaviour that sustain the autonomy of individual living things. There is no clear separation between a mind and a body. What actually exists is a "continuity between biological autonomy and sense-making" (Ramírez-Vizcaya & Froese, 2019, p. 4).

This idea of enactivism circles back to many themes already introduced. For example:

- The description of habits draws on multiple characteristics of complex systems (Chapter 3).
- The idea of extended habits resonates with sociocultural views of learning (Chapter 5). Claxton's metaphor of building learning power also draws on sociocultural theory—specifically the idea of "communities of practice", which emphasise that habit formation is essentially about the usual ways of working within the group (Claxton, 2012).
- Habits are not limited to human mind–bodies. When habits are viewed through a complexity lens, there is no binary between humans and other living things (Chapter 8).

All these ways of thinking about habits have practical implications for the classroom.

Building habits for complex systems thinking

Habit formation takes weeks and months, not hours and days.
(Claxton, 2012, p. 14)

At the time of writing this chapter in the second half of 2020 I was attempting to build a new habit. I needed to remember to use my phone to scan QR codes in shops and meeting places in case the COVID-19 tracers should need to alert me to a community outbreak of the virus that could potentially impact me. At first, I had to consciously think about it, and sometimes I forgot. I also felt a bit awkward—lots of other people were not taking any notice of the QR codes as they complacently went about life-as-normal. Gradually the process became more automatic and I stopped even thinking about what other people were doing. This has now become my "new normal" way of *being* in public spaces but it still feels fragile. If I stop doing it I am pretty sure the habit will disappear again.

I imagine that anyone who reads this could describe a similarly effortful habit change. How might these personal experiences be leveraged to create learning opportunities for students? One obvious point is that, at least initially, habits need to be brought to conscious awareness. Teachers need to help students notice and practise thinking habits. They also need to notice their own habits in the classroom (Claxton, 2008). Another point is that a habit needs to feel normal. As Claxton would say, it needs to be "the way we do things around here". That means teachers need to model the habits students are expected to build. Nelsen (2015) also stresses this point. The set of habits I introduce next was designed to support teachers to do these things in the classroom.

Habits of systems thinkers

The Waters Foundation in the USA has been working for some years on providing resources to support systems thinking.[37] Over time, their team of researchers has developed and refined a set of habits of systems thinkers that teachers and students can use as a resource during discussions. They describe 13 habits of a systems thinker, who:

- observes how elements within systems change over time, generating patterns and trends
- changes perspectives to increase understanding
- identifies the circular nature of complex cause-and-effect relationships
- considers short-term, long-term, and unintended consequences of actions
- recognises the impact of time delays when exploring cause and effect relationships
- seeks to understand the big picture
- recognises that a system's structure generates its behaviour
- checks results and changes actions if needed: "successive approximation"
- surfaces and tests assumptions
- considers an issue fully and resists the urge to come to a quick conclusion

---

37 https://waterscenterst.org/systems-thinking-tools-and-strategies/habits-of-a-systems-thinker/

- pays attention to accumulations and their rates of change
- makes meaningful connections within and between systems
- considers how mental models affect current reality and the future.

Notice that most of these habits imply that the learning action is happening *in a curriculum context*. Cause-and-effect relationships explain something specific (see Chapter 4). Assumptions are made and perspectives are explored in specific contexts (see Chapter 7). Accumulations and rates of change can be drawn as stock-and-flow diagrams in many different contexts (see Chapter 6). These habits are intended to be fostered across the curriculum as part and parcel of everyday learning conversations, with students of all ages (see for example Benson, 2020). That is how habits are formed.

Schools can buy illustrated card sets of these habits for classroom use. In Figure 12 I have chosen one card to illustrate, showing both the face side and the flip side. I chose this example because it supports metacognitive thinking as well as content thinking, and hence has interesting implications for specific conversations about habits themselves.

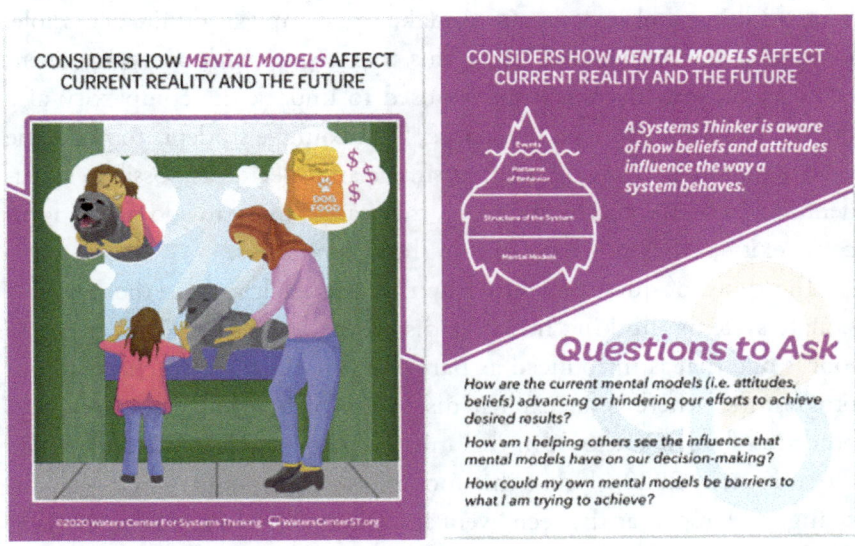

Figure 12. An example of one "Habits of a systems thinker" card, as developed by The Waters Foundation[38]

---

38 https://thinkingtoolsstudio.org/cards

The iceberg thinking tool shown on the flip side of the card in Figure 12 can be used for surfacing mental models deep under the surface of our day-to-day thinking. Gerber (2017) has written an interesting and accessible blog post on how to use iceberg as a systems-thinking tool.[39] The tip of the iceberg, above the water line, represents the event to be explored. His example is the familiar act of stopping at a red traffic light. Doing this is part of a pattern of driving behaviour (to which I would add, "and habits"). That's represented by the first of the underwater layers on the iceberg tool. The middle underwater layer represents the systemic structure that governs and shapes that behaviour: there are laws to obey, signals to notice, traffic officers to watch out for, how other drivers tend to behave, and so on. These aspects of the context help shape the behaviour of the individual. Finally, the deepest layer gets at the driver's mental model of how to behave in such events. Extending Gerber's example, we could ask whether the driver is risk averse or do they habitually try to run red lights? How do they perceive traffic rules in relation to themselves and others?

Even in this very simple example, there is a lot to unpack. The iceberg thinking tool is designed to make the complexity quite accessible. I found it interesting that using this tool begins with a specific event or problem, just like the sequence used to unpack the complexity of a pond ecosystem described in Chapter 5. Encourage students to make the "above the water" starting point as simple and specific as possible. If students begin with something too big, they can struggle to add meaningful complexity as they move down into the deeper layers.[40]

The types of questions posed by the deepest layer cue the compassionate systems thinking initiative discussed in Chapter 7. Indeed, this tool is one that is introduced as part of the toolbox for teachers in that programme. There is an explicit discussion of its uses in the resources produced for IBO schools in this initiative (International Baccalaureate Organization, 2020). Using this tool allows students to "grasp that for change to happen at the seen event level, systemic change is required at the lowest belief or 'mental model' stage" (p. 23). Understanding this dynamic interplay between beliefs and actions is an important insight

---

39 https://agsystemsthinking.net/2017/11/13/iceberg/

40 My thanks to Jane Drake for this tip, based on her observations of the tool in use.

when compassionate systems thinking is seen as an important educational goal.

If you visit the Thinking Tools Studio on the Waters Foundation website and flip all the cards, you'll find several other simple tools designed to make thinking visible to students. Stock-and-flow diagrams, change-over-time graphs, and the ladder of inference have all made an appearance in earlier chapters. All these tools are also outlined in the IBO resource just mentioned (International Baccalaureate Organization, 2020).

I looked for research about the effective use of these tools. A number of teachers have blogged about how they use them, but I didn't find any peer-reviewed journal reports. I did find a summary report that was self-published by The Waters Foundation (The Waters Foundation, n.d.). This summarises action research in many classrooms, over a number of years. The following quote gets to the essence of the findings. I chose it because it speaks to the concern that complex systems thinking might be seen as too hard for some students. The report makes the case that the conversations prompted by the habits cards can provide an accessible way for *all* students to contribute in the classroom and make meaningful learning gains:

> A systems thinking learning environment is motivating and engaging for even the most reluctant learner. Teachers report that the visual nature of the system thinking tools enables students to organize and express their thinking. The tools help motivate those children who tend to appear less involved, shy or reluctant to fully engage in learning activities. Teachers recognize that these children, just like their peers, are natural systems thinkers, as they readily make connections, embrace the big picture, and eagerly share new insights into the systems they're studying. (The Waters Foundation, n.d., p. 4)

Other useful "habits" resources

Many schools use the building learning power approach and resources (Claxton, 2012, 2018). The metaphor of split-screen thinking could also be used to complement the complex systems questions on the flip side of the Habits of Systems Thinkers cards. Thinking about the complexity inherent in a learning context is on one screen while the substantive content of the learning is on the other screen. Collaborative conversations, guided and supported by the teacher, bring both together.

I am also aware that a number of New Zealand schools already use the Habits of Mind approach and resources. Figure 13 presents a visual version of these habits. Notice that they describe more generic learning dispositions, with bundles of habits suggested by the small print.

Figure 13. Habits of Mind, as developed by the Habits of Mind Institute [41]

As I see it, this set of habits complements the habits of systems thinkers, but with one important gap. This set of habits has been developed and refined over many years, by researchers Art Costa and Bena Kallick. They envisage these habits as what intelligent people do when they are confronted with problems that they don't immediately know how to resolve (Costa & Kallick, 2008). There is considerable overlap with the habits of systems thinkers, but more explicit complexity concepts are not represented—this is the gap that I just mentioned.

When dealing with complex systems there are also several important caveats to the idea of "not knowing what to do". The first is that complex systems concepts are unfamiliar and can be counterintuitive to those

---

[41] https://www.habitsofmindinstitute.org/wp-content/uploads/2018/10/HabitsofTheMindChartv2.pdf

of us steeped in Western patterns of thought (see chapters 3, 4, and 8). Drawing on past knowledge may not be a good-enough guide, unless and until we have had multiple opportunities to learn about, and in, complex contexts. The habit of thinking in this way is itself a legitimate learning target. The second caveat concerns the assumption is that it is even possible, given sufficiently intelligent thought, to "know what to do". Complex systems can be unpredictable. Complex problems may not be amenable to resolution—fixing one thing can inadvertently cause new problems to arise. Learning to hold these uncertainties is not quite covered by any one of the habits of mind.

Building a varied repertoire of habit-building contexts

> *Pūtātara* encourages schools and teachers to create learning opportunities that expand learners' understanding of complex issues and take action for change. The resource is aligned to best practice in local curriculum design, education for global citizenship, and environmental education …[42]

Habits of Mind were developed by researching what people do when they don't know what to do. Nelsen's review of the nature of dispositions emphasises the importance of novel situations (Nelsen, 2015). If we want new habits to emerge, students need opportunities to explore many different contexts, using the doubts and uncertainties that might arise to inspire inquiry and help develop more flexible thinking habits (see also Chapter 11). Teachers face practical challenges, not least having enough time, for designing and resourcing rich inquiry topics that can meet this brief. I thought it would be useful at this point to introduce a curated collection of inquiry topics already designed with complexity in mind.

The Pūtātara website is a repository for resources designed to develop "learning opportunities that are place-based, inquiry-led, and focused on participation for change".[43] Notice the added bonus of the emphasis on place-based learning. Meaningfully introducing this type of pedagogy emerged as a significant challenge in Chapter 8. It is central to culturally

---

42 https://putatara.education.govt.nz/#/about
43 https://putatara.education.govt.nz/#/home

responsive teaching, and to developing students' sense of identity and belonging as well as their growing intuition for recognising the complex interconnectedness of everything in the world.

The resources on this site are organised into six inquiry pathways. I have copied and pasted the first paragraph of the overview of each pathway to show how strongly the themes from both chapters 7 and 8 come together to create many different inquiry possibilities:

1. **Tūrangawaewae: Where we stand, we listen** leads learners to uncover the histories of the local place in which they stand, and how these are reflected in their world today. It explores the major themes of time, continuity, change, and collective identity. It might uncover messy and layered histories of colonisation, deforestation, societal amnesia, and historical erasure. Equally, it might reveal the love and care we have for one another, for our elders, and for the place in which we stand.

2. **Tūrangawaewae: Being in and of this world** is a holistic exploration of personal and environmental wellbeing and how they are inextricably linked with one another. It leads learners to dive deep into their own identity and the ways in which their wairua is connected to (or disconnected from) their natural environment and landscape. This is a powerful foundation for exploring active citizenship in Aotearoa New Zealand.

3. **Kaitiakitanga: Our wonderful world** leads learners on a journey of both curiosity and reverence as they inquire and learn about the natural legacy we have inherited in Aotearoa New Zealand. It encourages kinship, responsibility, and action on environmental and social justice issues and provides tangible ways for schools to contribute positively to the regeneration of our collective mauri.

4. **Kaitiakitanga: Think globally, act locally** leads learners to consider themselves as global citizens. It focuses on solutions to climate change, a rich context in which learners can develop critical-thinking and problem-solving skills that support action and participation in the real world.

5. **Whakapuāwai: The future is circular** examines the opportunities that exist to transform our ways of living. It explores ideas about linear economies and globalisation, considering the

impact many products have on the planet and the people that work to create them.

6. **Whakapuāwai: Future horizons** explores possibilities for a sustainable world, considering economies, trade and sustainability, and the nature of technology and innovation. It examines how humans can live in concert with the planet while residing in cities, which now house the vast majority of the global population.

I hope you will explore the website for yourself, and check out the supporting resources such as short videos suitable for classroom use. Each inquiry pathway is designed to integrate different learning areas of *NZC*. Next, I turn in Chapter 10 to critical curriculum-design challenges.

Chapter 9 reading guide

Dispositions for complex systems thinking can be purposefully fostered by supporting students to develop relevant thinking habits. The relationship between dispositions and habits is itself complex. For teaching purposes, dispositions can be seen as "bundles of habits" which can be strengthened individually and collectively. Advice about how to do this broadly agrees with the pedagogical suggestions discussed in previous chapters. Practising in many different contexts is important and so is collective metacognitive reflection.

1. The Chapter 8 reading guide explored habits of Western thinking that can be hard to break. This chapter extends that discussion by problematising the idea of a separation between our minds and our bodies (called the mind/body binary). In what ways is this idea deeply embedded in Western assumptions about learning and traditional pedagogical practices? Why do such assumptions need to be revisited when we think about educating for dispositional change?

2. At the start of this chapter I acknowledge my own tendency to take the dispositional aspects of competencies for granted. In the past, I acted as if I thought that teaching in ways that make space for developing students' competencies is enough for envisaged changes to follow. In what ways does this assumption now seem problematic? Does the idea of developing dispositions as "bundles of habits" seem practical to you? Why or why not?

3. Can you recall the "4Es" of habits? Does this categorisation make sense to you? How does each of these terms reflect ideas about complexity introduced in Chapter 3?

4. Which, if any, of the resources and pedagogies introduced in this chapter have you thought about or do you already use? Which do you think might be most useful for fostering dispositions for complex systems thinking in your context? Why do you think this? If you are discussing this reading guide as a team, you could consider allocating different resources to different people to explore and then report back on. Alternatively, you might like to explore one of the interactive online resources as a group.

Chapter 10
Curriculum-design considerations

The school curriculum could be envisaged as a rich mosaic of complementary pieces. The different types of contributions made by traditional and newer subjects have the potential to build to an interwoven whole, where the sum is more than the parts, and opportunities to strengthen systems thinking abound. However, collaborative, co-ordinated design is needed to achieve this vision.

Looking back over the previous chapters, a number of clear curriculum messages can be distilled from the various challenges explored. This chapter begins with a brief synthesis of those messages, before moving on to implications for curriculum design.

Five clear curriculum messages
From a synthesis of the previous chapters, five clear curriculum messages stand out. Like so much about complexity, they are deceptively simple.

- **Learning for complex systems thinking has relevance for** all learners

    The examples included in previous chapters span all levels of the school system, from very young students, through middle school, senior secondary subjects, and tertiary-level studies. There is a clear message that the visual tools designed to support complex systems thinking make the envisaged learning accessible for students of differing abilities and ages (Benson, 2020), including those with special learning needs (The Waters Foundation, n.d.). This inclusive message resonates with an example from my personal experience—some IBO teachers reported greatly increased engagement from underachieving students when they introduced stock-and-flow diagrams as a way to explore complex dilemmas.

- **Content knowledge is important but how it is introduced might need to be adapted**

   Many complex phenomena are already included in the school curriculum. Examples include various physical and biological systems in the sciences, and complex social-sciences issues such as inequality. However students will not become systems thinkers simply by *learning about* specific examples of complex systems—complexity concepts need to be introduced in ways that build bridges between highly theoretical systems knowledge and more traditional curriculum content. Traditional reductionist pedagogies work against developing complex systems thinking. Therefore, content needs to be introduced in ways that maintain strong connections between the whole and the parts, and that locate students as part of the systems they learn about, not outside them looking in. Complexity learning cannot be delegated to one or two subjects. It needs to be a focus in multiple learning contexts if we want students to build their dispositions to be and become complex systems thinkers.

- **Many complex issues are not neatly confined within the bounds of traditional curriculum subjects**

   This message implies that some form of curriculum integration is needed. Different researchers envisage somewhat different ways of bringing subjects together. An *interdisciplinary* curriculum might bring discrete topics from different learning areas together in a unit where learning about complexity is the unifying topic (e.g., Heinrich & Kupers, 2019). *Multidisciplinary* approaches might foreground one or more specific capabilities associated with systems thinking, applying these in different disciplinary contexts. I'll suggest some examples shortly. *Transdisciplinary* approaches foreground *student* inquiry into a complex issue, drawing on disciplines as necessary. The inquiry pathways introduced at the end of Chapter 9 imply this type of integration. Whichever approach is used, an important curriculum-design challenge will be to meaningfully weave traditionally separate subjects together.

- **Building a rich local curriculum is important**

    This theme emerged strongly in the discussion of parallels between indigenous knowledge systems and complexity sciences, but it is also apparent when complexity is discussed in the sustainability literature more generally (e.g., Bernier, 2018). Students need to build meaningful connections to the natural environment that surrounds and sustains their communities, and to become more aware of the impact of humans on the natural systems that sustain life on the planet. It is not enough to *learn about* these things—being immersed within systems helps build embodied, intuitive ways of knowing that decentre individual interests and help build a more collective ethic of care.

- **Learning about complexity is important for citizenship**

    There is a clear and consistent message about purposes for learning running through the various papers reviewed for this book. If young people are going to be and become responsible citizens, they need opportunities to get to grips with the idea of complexity, to understand the dynamics of complex systems, and to develop the habit of seeking out complexity, along with thinking in systems terms when confronting problems and challenges. Young people need to be able to explore challenges from multiple perspectives, tolerate uncertainty, and to develop the compassion necessary to actually care about outcomes for everyone, and for the planet as a whole. This is a huge agenda that can only be achieved if relevant experiences are included in the curriculum across the years of school, and within multiple contexts, including different subjects. That is the curriculum challenge to which I now turn.

Translating good intentions into specific curriculum design

Learning for citizenship, broadly speaking, is signalled as important in the overarching framework or "front half" of *NZC*. It is implicit in the vision of educating young people to be *"confident, connected, actively involved lifelong learners"* (p. 8) and in the inclusion of key competencies, especially *participating and contributing* (p. 13), and in values statements about *equity, community and participation, ecological sustainability, integrity*, and *respect* for themselves, others, and human rights (p. 10). The citizenship message is explicit in several of the pithy purpose statements for the different learning areas. For example, students learn science "so that they can participate as critical, informed and responsible citizens, in a society in which science plays a significant role" (p. 17). The statement for the social sciences is very similar. Learning technology helps young people to become "discerning consumers who will make a difference in the world" (p. 17). They learn languages in part to "explore different world views in relation to their own" (p. 17). This high-level sense of purpose is also implicit in the extended versions of the purpose statements for the other learning areas (pp. 18–33).

Citizenship as an overarching purpose for learning is also central to curriculum models developed elsewhere, including by major international organisations. For example the OECD's "learning compass" initiative asks:

> How can we prepare students for jobs that have not yet been created, to tackle societal challenges that we cannot yet imagine, and to use technologies that have not yet been invented? How can we equip them to thrive in an interconnected world where they need to understand and appreciate different perspectives and worldviews, interact respectfully with others, and take responsible action toward sustainability and collective well-being? (OECD, 2019, p. 5)

This sort of rhetorical curriculum thinking needs to be translated into actual curriculum design. That is easier said than done but many nations are exploring the challenge. One recent UNESCO report says there has been a strong *reorientation* of many national curriculum documents towards a focus on sets of competences or capabilities (Marope et al., n.d). These *can* foster dispositions to be and become complex systems thinkers, provided that is understood to be one of their roles, and they are presented as coherent and consistent curriculum entities. The report suggests that curriculum documents do not always convey clear messages about the purpose of components such as key competencies and I suspect that many New Zealand teachers would agree. The report also highlights the challenge that competences/capabilities are too often seen to be in competition with traditional curriculum content. A binary either/or view needs to be replaced with a more nuanced view about what must change and what can stay the same:

> Curricula are not only about change, they are also about stability. Most particularly, the stability of core functions of education such as the facilitation of foundational/enabling competences like basic literacy, knowing how to learn, and mastery of fundamental disciplines like language, sciences, and mathematics. Curricula reforms must mitigate the risk of crowding out these core functions even through education and learning systems' efforts to be responsive. They must strike a delicate balance between change and stability (Marope et al., n.d, p. 16).

Given the two-part structure of *NZC*, New Zealand teachers face the specific challenge of bringing together the high-level features that signal re-focused purposes for learning, and the more traditional learning outcomes that provide for continuity. However it is not self-evident how this bringing together should happen. One suggestion is that key competencies are positioned as "ideas to think with". Instead of being added on top of content, they should prompt a rethinking of the *purposes for learning* envisaged when working with disciplinary content (e.g., Hipkins, 2019b; Hipkins & Bull, 2015). My colleagues and I have used the metaphor of "weaving" the curriculum components together, again with clear purposes in mind (Hipkins, 2019b). My plan in what follows is to reframe the weaving challenge to suggest that various traditional school learning experiences could be treated a bit like a mosaic of pieces that might make different, complementary contributions towards the development of complex systems thinking.

How subjects might contribute to complex systems thinking

Building knowledge, skills and dispositions (i.e., competences or capabilities) for complex systems thinking is seen as essential for citizenship. With this clear purpose in mind, it follows that the next curriculum design step is to find ways to bring systems-thinking capabilities and traditional content together. I'm going to tackle that challenge from a somewhat oblique angle. My plan is to explore how subjects, as they are traditionally understood, might make different types of contributions to developing systems thinking.

Conceptual learning

A recent report from the Education Review Office (ERO) identifies "a mistaken belief among teachers that KCs risk replacing knowledge learning". They say that, in fact "the reverse is true. The KCs and knowledge complement one another" (Education Review Office, 2019, p. 15). It seems likely that this belief stems from either/or (binary) ways of thinking that are hard to dispel. At the risk of labouring the point, many subjects already include content related to complex systems and this conceptual learning is still important. However new opportunities arise when complexity is made a specific learning focus. Chapters 5 and 6 include the science example of teaching particle theory with a specific focus on the agent-based behaviour of individual particles. The research suggests that explicit attention to this complexity concept could help move students past the developmental tendency to think in anthropomorphic terms, via which they see particle behaviour as intentional (Samon & Levy, 2020).

Not every concept traditionally included in the curriculum can be explored in the depth implied by the research on the agent-based behaviour of particles. Some content reduction will inevitably be needed, and this should be done in a principled way. Here are two different ideas for making principled decisions about content reduction.

- **Focus on concepts that create "hubs" in webs of interconnections:** The particulate nature of matter is a *powerful idea* because it forms a central hub in a vast network of knowledge (Holbert, 2016). Thinking about interconnectedness is itself a complex systems idea and so I see this as one useful principle for making content choices.

- **Focus on "big understandings":** In his book *Futurewise*, David Perkins suggests that content choices should focus on "big understandings". He offers four ways in which ideas might be considered to be big. Such ideas: build new insights for students; enable action—students can do something with the learning; help develop ethical mindsets and conduct; and are likely to afford multiple opportunities to use the knowledge gained (Perkins, 2014, p. 52). This is a pretty good set of principles too.

Chapter 3 outlined conceptual knowledge about complex systems that also needs to be explicitly taught. Curriculum designers need to plan for when and how that will happen. I imagine it would be counterproductive if every individual subject traversed the same ground, even with the contextual differences afforded by discipline-specific examples. But nor can a direct focus on complexity concepts be allowed to "fall between the cracks" if the curriculum does not allocate specific space to this important learning. This challenge points to the importance of *collaborative and co-ordinated* curriculum design—at least at the high-level. Siloed planning within individual subjects will not achieve the necessary co-ordination.

I wonder if there are additional concepts that are not necessarily taught to most students at the moment, but that have big implications for understanding the dynamics of complexity. My concern here is situated in the senior secondary school in particular, where students have very different learning experiences depending on the combination of subjects they choose. As one example, several of the papers I read while writing this book mentioned a concept called the tragedy of the commons (Heinrich & Kupers, 2019; Heke et al., 2019). Heke et al. describe this as "a well-known systems structure in which competition for a scarce and common [shared] resource leads to all losing out" (p. 27). In their case the scarce resource was coaching expertise needed for waka ama teams in Tolaga Bay. They developed potential solutions to this problem by using causal-loop mapping to move beyond traditional either/or thinking about who would be coached and who would miss out. Working as a team, researchers and local leaders developed a broader set of culturally embedded responses that could lead to sustained opportunities for all.

A number of recent online commentaries discuss the impact of the concept of the tragedy of the commons, 50 years after it was first developed by Garritt Hardin. As well as a clear theme that the tragedy lies in traditional either/or ways of thinking about scarcity and ownership, there is an interesting thread to the commentary about the importance of intellectual commons, especially in the complex environment of online sharing (e.g., Boyle, 2018). It seems to me that this is something that all students should understand, regardless of the combination of subjects they choose. No doubt there are other concepts that could and should be debated as potential candidates for inclusion in any curriculum that intends to build capabilities for complex systems thinking.

Finally, any topic in a traditional learning area could potentially provide opportunities to develop one of more of the systems thinking habits outlined in the previous chapter. As just one example, Benson (2020) describes several ways in which young children drew systems thinking conclusions from the story of the three little pigs. These are direct quotes from two children:

> The story of the three little pigs is like a fix that fails because pigs with houses of straw or sticks never really solved their problem of needing shelter that protected them from the wolf. [Fixes that fail are described as "systems archetypes" in Benson's paper—so here is another concept that is widely used in systems thinking contexts.]

> Sometimes you need to spend more time and resources if you really want to solve a problem for good, like the pig who build his house out of bricks. (Benson, 2020, p. 3)

Here the children are seeking out systems complexities in the fantastical context of a popular fable. They have obviously learnt to transfer their knowledge about "fixes that fail" to new contexts. Reading these comments, made by children who were evidently highly engaged, reminded me of the Lifelong Literacy research several of my colleagues completed some years ago (Twist & McDowall, 2010). In that project the researchers supported middle-primary teachers to weave key competencies into their daily reading programmes. The rich conversations that emerged seem similar to those described by Benson. The approach taken in the Lifelong Literacy project included inviting rich conversational links between children's lived experiences and the books they were reading in class. I think this idea could be readily adapted to include a complex systems thinking component in a classroom reading programme.

Disciplinary-inquiry practices

Much has already been said about the centrality of inquiry practices to the exploration of complex systems. Just to recap:

- Chapter 5 identifies ways that discipline-based inquiry pedagogies can be adapted so that complex systems thinking is fostered alongside knowledge of the system being explored (e.g., Hmelo-Silver et al., 2017).
- Chapter 6 introduces a range of e-inquiry tools, and outlines principles for recognising tools that will be effective for supporting inquiries that draw on the sorts of practices researchers themselves would use (Dorsey, 2020).
- Chapter 7 includes a section on effective use of modelling as an inquiry process. Models can be used to: stimulate curiosity and prompt question asking; decide what data are relevant to collect; check assumptions and different ways of seeing; explore trade-offs and efficiencies; and so on (Epstein, 2008).

Since each discipline has its own distinctive inquiry practices, it follows that students should be offered opportunities to investigate complex systems from a range of disciplinary perspectives, and in a range of subjects. Chapter 5 also discusses the importance of fostering metacognitive awareness of the thinking skills being used in any complex scenario. This is itself a type of inquiry and I now suggest an interesting additional possibility. If the necessary curriculum co-ordination is achieved, metacognitive reflection could also encompass the different approaches used in different subjects to investigate complex systems. In this scenario, inquiry might become the common element in a multidisciplinary approach to curriculum integration.

Pattern-seeking might be another candidate for a multidisciplinary approach to curriculum integration. A number of the habits of mind outlined in Chapter 9 imply attention to patterns of one sort or another. Again, inquiries of this type are already included in many subjects, and in some cases they already have explicit links to complexity ideas. For example, mathematics includes topics such as fractal geometry, where iterations of basic patterns can lead to ever-more complex structures. The example shown in Figure 14 is from the Figure it Out resource series, and is set at *NZC* Level 4.

Teaching for complex systems thinking

Science also already includes pattern-seeking exercises that could be adapted to include a more deliberate focus on complexity. Learning to notice regularities and changes in natural systems was discussed in Chapter 8, which introduced the deceptively simple idea of a walking curriculum. Keepers of complex knowledge about local ecosystems are adept at intuitive pattern thinking that takes years to develop, and hence requires modelling and practice. Pattern-seeking, or pattern sniffing, has also been identified as an important habit of mind to foster in future engineers (Lucas & Hanson, 2016) and would no doubt be seen as an important habit to foster for other career pathways.

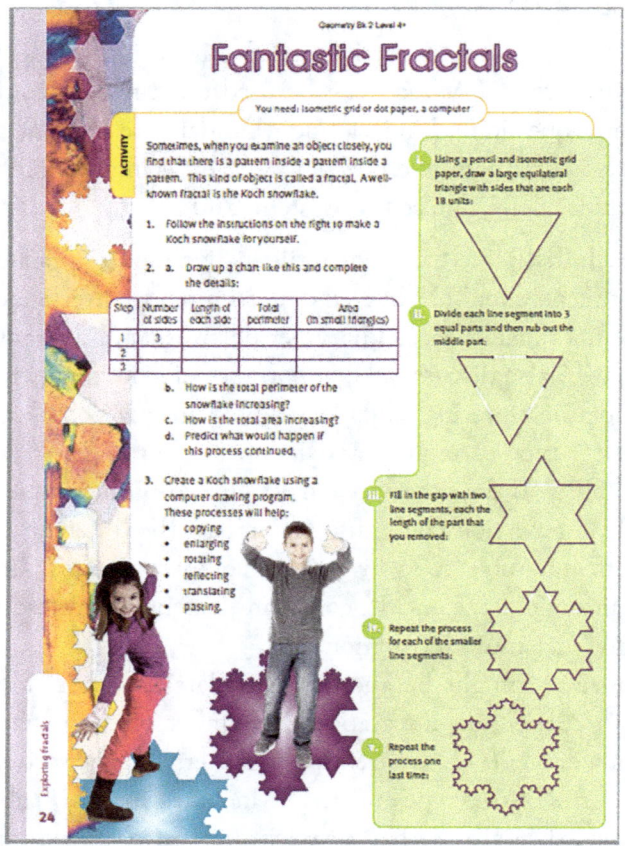

Figure 14. An example of a simple inquiry into fractal patterns in mathematics[44]

---

44 Source: NZ Maths website: https://nzmaths.co.nz/resource/fantastic-fractals

Close attention to patterns in texts, and to the "particularities and complexities of human experience" (p. 154) are two important aspects of literary inquiry in English (Sumara, 2002). Sumara explains the importance of supporting students to develop a "focal practice" that helps them to make sense of their life experiences. He says that introspection is not enough—we need access to the insights of great writers who have struggled to find ways to interrupt the familiarity of their own experiences, so that they can give us different ways to think about issues and challenges we might face. He says that practices that help students attend closely to patterns in texts might include close reading, memorising pieces, and copying syntax. He notes that none of these are things most students are likely to do without support and encouragement from a skilled teacher.

Subject-specific creative thinking
Chapter 5 noted that cause-and-effect thinking about complexity requires "a disciplined blend of critical and creative thinking: critical because [students] need to look for hidden connections; creative because they need to look beyond the obvious to find non-linear links and interactions; and disciplined because this is not a case of anything goes" (Drake et al., 2017, p. 32). Notice that critical thinking is listed before creative thinking here. I suspect this reflects their priority order in many subjects. My focus now is those subjects where the intention to foster creative thinking is explicitly in the foreground. The arts are an obvious candidate and I found an interesting discussion of one way they might contribute to the development of more complex thought patterns.

Silva Pacheco (2020) draws on recent neuroscience research to explain that highly creative people can co-activate three areas of the brain that usually act separately from one another. These areas are our "default network (used when the brain is at rest or imagining), salience network (which is used to discern about the importance of objects) and central executive network (which is activated in decision-making)" (p. 245). He says that when students bring critical and creative thinking together in *collective art-making activities*, they too, are given the opportunity to strengthen their ability to co-activate these brain areas and hence strengthen their complex-thinking capabilities.

Note that I have not included the word "systems" here because Silva Pacheco is concerned with the complexity of thought per se (see Chapter

14), but I can see interesting parallels with the idea of "sensing in" from the compassionate systems thinking model (Chapter 7). For me, this focus on neural functioning was a useful reminder that fostering dispositions to be and become complex systems thinkers has a *biological* aspect as well as more familiar psychological and social considerations. Silva Pacheco's message is clear: the arts can make a strong contribution to neural development that strengthens complex-thinking capabilities.

I don't want to imply that the arts are the *only* subjects that can do these things—other areas of the curriculum also provide rich opportunities for foregrounding creative thought. Literary writing, in English, is another candidate. Like artists, writers devise strategies to disrupt everyday ways of seeing and thinking, so that new insights might emerge (Sumara, 2002).

There are provisos about how creative learning experiences are shaped. Silva Pacheco is critical of traditional arts pedagogy which focuses on individual acts of production. He emphasises the importance of critique which fosters greater awareness of both individual and collective meaning-making and demands that students tolerate uncertainty and ambiguity during the creative process. An ethnographic study of the learning that took place during two collaborative arts-based community youth projects came to similar conclusions (Soep, 2006). Soep says that "critique erupts as a resource for learning" (p. 767) when students work together to produce an artwork in conditions of mutual and interactively sustained accountability, to meet a high-stakes goal. Interestingly, she also notes that such learning is almost inevitably interdisciplinary, because a complex production cannot be contained within single discipline boundaries.

The contribution of "practical" subjects

Now I turn to the subjects that are sometimes seen as the poor relations in an academic curriculum. Secondary students on the "well-lit" pathway to university often face timetabling impediments to mixing practical and academic subjects (Hipkins & Vaughan, 2019). It seems that pejorative thinking about "vocational" subjects is deeply embedded in our schooling system—right down in the deepest layer of the iceberg (Chapter 9). In complex systems terms, we might say that a bias against practical subjects is a long-range memory effect, embedded in both the informational and material legacies of our education system (Chapter 3). In the spirit of pushing for a shift in this dynamic, I turn now to the potential contribution of practical subjects to complex systems thinking.

> Making is a literacy—a way of reading the world as a collection of resources and materials to be composed, repurposed and rearranged. Making is "what if?" and "why not?"—of taking responsibility for challenges and obstacles faced by oneself and one's community and enacting solutions. (Holbert, 2016, p. 2)

This quote comes from a commentary on the ideas behind the maker movement. Maker spaces in schools are a comparatively recent phenomenon, but subjects with a strong "making" component have been included in the curriculum for many years. The possibilities include various technology subjects, along with home economics which has a quite different emphasis to food technology. Digital technologies are a recent curriculum addition—one with strong links to the development of complex systems thinking. Using computer simulations to undertake system modelling requires students to engage in both systems thinking and computational thinking simultaneously (Damelin et al., 2019). Playful building and testing of systems algorithms can take place in both "unplugged" and digital contexts, with peers testing one another's design and providing critique that leads to debugging.[45]

Opportunities for students to make things with their hands can bring together embodied and intuitive knowing with powerful theoretical ideas that might otherwise be inaccessible to students (Holbert, 2016). A review of the impact of the maker movement also suggests that the act of making encourages learning dispositions "by nurturing curiosity, exploration, and collaboration that comes with experimenting" (Deloitte, n.d., p. 19). The Deloitte report also notes that "the hands-on experience of tinkering, failing, and rapidly iterating allows learners to focus not on the physical outcome or product created, but rather on the actual creation process. The process is where meaningful learning occurs" (Deloitte, n.d., p. 19). When students are exposed to rich interconnections between materials and products they can gain new perspectives on their own relationship with the material world of everyday products. For me, all these

---

45 Stephen Ross, a digital technology adviser from the University of Waikato, provided these suggestions.

comments echo the idea of "sensing in" from the compassionate systems thinking initiative (Chapter 7). In short, making locates students inside familiar material systems of everyday life.

There is an important caveat about the learning conditions that can generate these benefits—one that echoes warning made in the context of the arts (Silva Pacheco, 2020). Making demands both creative and critical thinking, provided that the learning opportunities allow for open-ended experimentation:

> Quick and simple activities that result in highly polished products may in fact be detrimental to putting learners in contact with powerful ideas … it is in the mistakes, the restarts, and the tinkering where we often have the opportunity to reflect and make sense of encountered concepts … it is in the play and exploration that the ideas become our own (Holbert, 2016, p. 4).

This comment also resonates with the argument in Chapter 4 that reductionist pedagogies are likely to work against the development of complex systems thinking. Just as students need iterative opportunities to explore complex relationships between parts and wholes, so they need opportunities to directly experience the consequences of their design and construction choices when making complex artefacts. Note that I don't mean to imply that the envisaged product should be complicated—complex systems can in fact appear deceptively simple (see Chapter 4).

Integration as a both/and design challenge

The suggestions I have made in this chapter are not intended to provide an exhaustive summary of the potential contribution of different curriculum subjects. I have not, for example, circled back to ideas about using physical games to build complexity concepts (see chapters 3 and 5). Nor have I included newer subjects that already come at least partly integrated, for example media studies, environmental education, or agribusiness.[46] My aim was simply to think through different ways that familiar subjects might make complementary contributions to an integrated curriculum, given the clear message that integration is important for complex systems learning.

---

46 See https://www.agribusiness.school.nz/ if the idea of agribusiness as a school subject is new to you

I am very aware that any form of curriculum integration is challenging to do well. Some colleagues and I recently worked with teachers from a range of school settings who had volunteered to share their curriculum-integration experiences with us. These teachers had ambitious curriculum goals and they were willing to think about curriculum design in new ways, but they also encountered difficult trade-offs as they attempted to balance opportunities for students to: take greater agency for their learning; learn "just-in-time" content knowledge; and take part in discipline-specific inquiry practices. The concluding section of the report commented:

> We want all students to see learning as relevant, engaging, and connected, and we want all students to develop conceptually and have opportunities to build knowledge, not just acquire and re-present information. This capacity opens doors to life chances: it enables students to design their own social futures and is necessary to solve the complex issues of our time. However, meeting the dual goal of student agency and knowledge building is a challenging undertaking for teachers even in the context of *single*-subject teaching (McDowall & Hipkins, 2019, p. 56, emphasis in the original).

The final sentence of this quote implies that doing these things will be even more challenging when two or more subjects are involved. This challenge feels huge, but we need to start somewhere or nothing will change. In the Kiwi spirit of "giving it a go" I have laid out a range of suggestions for drawing on different disciplines in ways that complement one another when fostering systems thinking is a unifying learning goal. There is no one right way to bring the pieces together—I think of them as being like the tiles in a mosaic. Whichever way the overall picture gets arranged it must make sense to those who will work with the design.

Box 11. Learning in Depth: A different way for students to experience curriculum integration

Teachers in the curriculum-integration project struggled to balance student agency and coverage of important curriculum content (McDowall & Hipkins, 2019). The curriculum initiative I am about to outline avoids this challenge because it takes up minimal formal classroom time—about 1 hour per week.

Most students are motivated to pursue their chosen topic in their own time, developing deep curriculum knowledge as they do so.

Learning in Depth is a pedagogical initiative developed by Kieran Egan from Simon Fraser University in Canada (Egan, 2010). (The same department developed the walking curriculum initiative outlined in Chapter 8.) On the website dedicated to the initiative the basic strategy is explained as follows:

> Each child is given a particular topic to learn about through her or his whole school career, in addition to the usual curriculum, and builds a personal portfolio on the topic. To the surprise of many, children usually take to the program with great enthusiasm, and within a few months LiD begins to transform their experience as learners. The program usually takes about an hour a week, with the students working outside school time increasingly. (Learning in Depth website)[47]

Students start with an age-appropriate topic. "Apples", "railways", "the circus" and "the solar system" are indicative suggestions. From simple beginnings, and with regular teacher support, they continue to build a portfolio of knowledge that becomes both broader and deeper across their years of schooling. The more they come to know about their topic, the more often new questions arise. When pursuing those questions, curriculum integration inevitably occurs. With encouragement, new conceptual learning can be incorporated into the individual network each student continues to build, encouraging transfer, and consolidating meaningful learning.

This initiative applies the insight that people who have taken the time to build deep expertise in one area are more likely to be aware of what they don't know when they encounter another area of personal interest or importance. The team who work with the initiative say that it transforms students' view of themselves as learners. It also seems to me that it builds very strong dispositional and knowledge foundations for complex systems thinking. For one thing, knowledge must become more interconnected as a portfolio grows. For another, adding new layers must help build new perspectives and insights, especially when students are supported to reflect on the new meanings they are making of their topic. Being aware that there is always more to know, along with learning to tolerate uncertainty, helps build dispositions for complex systems thinking (see Chapter 9).

---

47 http://ierg.ca/LID/

Collaborative, co-ordinated curriculum design

This chapter makes the case that curriculum design needs to be collaborative and co-ordinated, both horizontally and vertically, if complex systems thinking is seen as an important outcome of schooling. I say this because:

Space must be created to teach complex systems concepts: these are likely to fall between the cracks if they are no one's specific responsibility. The risk is more evident at the secondary level, but I foresee a challenge at the primary-school level too: when should concepts best be introduced? This is essentially a question about development and progression, and I address it in Chapter 13.

Meaningful curriculum integration requires clarity about the contribution made by each traditional learning area: my mosaic idea is all very well in principle. But a collection of pieces won't become a clear picture without attention and effort. Since different teachers have creative strengths and expertise in different subjects, collaboration is key to achieving a coherent, rich curriculum design.

Building a rich local curriculum implies collaboration beyond the school: I will delve further into this point in the next chapter. Meanwhile, the metaphor of two-eyed seeing (Chapter 9) could provide a novel way of thinking about partnerships with others in the community. Teachers bring a strong educative perspective and expertise while others in the community bring deep local knowledge both of the surrounding environment and systems, and of learners as members of local communities and families.

Collaboration between schools helps achieve continuity of learning experiences: early implementation of the key competencies in New Zealand schools showed the benefits of having a common language across a school when introducing pedagogical innovation(s) (Cowie et al., 2011). New teaching practices such as the encouragement of "it depends" thinking and the fostering of systems-thinking habits will also need language and routines that become part of "the way we do things here" (Claxton, 2012). It would be very disruptive and confusing if such practices were to be fostered in a primary school (say) but discouraged by new teachers when students move to a different school.

This implies a need for alignment of pedagogy across the schools in a community. In the same vein, Learning in Depth is a deceptively simple idea, but for most students, putting it into practice across all the years of their schooling would require co-ordination between schools. Teachers at all levels need to understand the trajectory of students' personal learning journeys, and how the learning that they foster fits within the whole.

Paying attention to the taken-for-granted

Quite late in the process of writing this book I read a paper that had just been published in *The Curriculum Journal*. This paper really got me thinking about what I had been taking for granted in this chapter and to some extent in all the previous chapters. Looking back, *contexts* for learning are implicated at every turn. I have said that we could think about traditional subjects as contexts for complex systems learning but otherwise the idea of choosing productive contexts lurks, undeveloped, behind the scenes. And so I turn now to a discussion the importance of contextual choices and explore some possibilities.

Chapter 10 reading guide

The school curriculum could be envisaged as a rich mosaic of complementary pieces. The different types of contributions made by traditional and newer subjects have the potential to build to an interwoven whole, where the sum is more than the parts, and opportunities to strengthen systems thinking abound. However, collaborative, co-ordinated design is needed to achieve this vision.

This chapter has a different structure and focus to the earlier chapters. It does not address ideas about complexity directly. Instead it invites readers to consider the implications for curriculum design if learning to think in complex systems terms is taken seriously as a goal of learning. Another difference is that some decisions and actions considered in this chapter cannot be undertaken by individual teachers alone. They depend on the actual curriculum structure and collective curriculum enactment across the school.

1. What is your reaction to the idea that the school curriculum could be rebuilt (from traditional pieces) as a rich mosaic that explicitly reflects high-level curriculum goals, including but not limited to complex systems thinking for informed and active citizenship? What in-principle opportunities and barriers do you see?

2. Between drafting this chapter and writing this guide, New Zealand's Ministry of Education announced a "curriculum refresh". It is too early for me to draw any conclusions on how this might impact the arguments laid out but I invite readers to do so once the refresh has been completed. How have the high-level signals in *NZC* outlined on page 124 shifted, if at all? Would you say that changes make the case for explicitly including complex systems thinking in the curriculum stronger, weaker, or about the same? (Readers from other nations could consider signals about purposes for learning in relation to the national or state-level curriculum that guides their work.)

3. Does your school have a culture of collaborative curriculum planning? If not, what could you do to support a wider conversation between teachers and/or teaching teams? In your own classroom, within the limitations of the existing curriculum structures, could you make changes to the enacted curriculum that open up opportunities for complex systems thinking? If so, what might these changes look like?

Chapter 11
Choosing contexts that support the development of complexity thinking

Strategically chosen learning contexts can open up opportunities to foster awareness, knowledge, and dispositions that strengthen complex systems thinking. Features of such contexts are likely to be some combination of: real-world issues and challenges that have meaning for students' lives and that bring different learning areas together; opportunities for collaboration with members of the wider local community and to take action to address an aspect of a complex issue; and/or opportunities to learn about, and where possible contribute to, complex questions of current researcher interest. The examples included in this chapter show that these types of contexts can be accessible to students of all ages, with appropriate support.

> The content *and the contexts* that we choose to explore will either invite or inhibit complexity. We need rich, relevant content that students can immerse themselves into in order to explore connections, to be challenged by perspectives and *experience the ambiguity that is at the heart of knowledge making*. (Drake et al., 2017, p. 33, emphasis added)

It's a bit ironic that I came late to the realisation that I needed to include a specific chapter about contexts in this book. I first became interested in the idea of "teaching in context" when New Zealand moved to a model of outcomes-based curricula in the 1990s (Hipkins & Arcus,

1997). Somewhere over the course of the intervening years, this focus on contexts faded into the background. Maybe this happened because I had seen a lot of trivial applications of the idea. What I saw was often little more than "candy wrapping"—essentially teaching traditional content but wrapping a context around it to provide a veneer of applicability that might engage students.[48] Being more critically aware of the role that contexts play in learning won't necessarily prevent this from happening of course. Nevertheless, the choice of generative contexts for learning is an aspect of curriculum-making that requires careful thought and purposeful choices.

Curriculum experts from Victoria University of Wellington recently published the results of a retrospective analysis of two previously completed projects set in social-science contexts, and funded by the TLRI.[49] Both projects focused on learning experiences that were intended to transform an aspect of students' capabilities for informed citizenship. Mark Sheehan and his team investigated the development of senior secondary students' historical thinking and historical consciousness.[50] Bronwyn Wood and her team investigated how students undertook critical social inquiry that culminated with taking action of some sort.[51] Neither of these projects had a direct focus on complexity thinking, but I can see strong parallels with some of the initiatives introduced earlier in the book. They are visible in this quote, for example:

> The transformative aspects of disciplinary learning in social studies, therefore, involve a commitment to critical and multiperspectivity understandings of social issues, alongside active civic engagement strategies that seek to transform oppression and create a more just and sustainable society. (Wood & Sheehan, 2020, p. 7)

---

48 I worry that this could also be the fate of well-meaning calls to weave elements of mātauranga Māori into the curriculum unless teachers have opportunities to embark on critical reflective conversations about different knowledge systems, as signalled in the Chapter 8 reading guide.

49 Teaching and Learning Research Initiative: http://www.tlri.org.nz/tlri-research

50 http://www.tlri.org.nz/tlri-research/research-completed/school-sector/%E2%80%98thinking-historically%E2%80%99-role-ncea-research-projects

51 http://www.tlri.org.nz/tlri-research/research-completed/school-sector/creating-active-citizens-interpreting-implementing

The need for critical perspective-taking, in combination with a focus on social justice, has strong parallels to the compassionate systems thinking initiative (Chapter 7). The overarching commitment to strengthening students' capabilities for critical citizenship in these TLRI projects is another strong parallel, as is the emphasis on sustainability. Among other findings from their synthesis, Wood and Sheehan argue that some teachers chose, or guided students to choose, topics and contexts that more readily allowed students to demonstrate their growing capabilities. By contrast, if teachers allowed open choice, some students chose inquiry contexts that constrained their ability to strengthen and demonstrate the capabilities being assessed. This then meant that they could not display the depth of capability development needed to do well in their NCEA assessments. To illustrate, in the historical-thinking project some students "engaged in substantive topics such as the experience of women and children in World War II, others studied the history of surfing or hip hop" (Wood & Sheehan, 2020, p. 10). Many teachers avoided the topic of New Zealand's difficult colonial history altogether, even though significance to New Zealand was one of the aspects students were required to consider in their chosen topic.[52] In the social-action project two male students chose to explore the right to wear facial hair at school. "Whilst passionate—and exhibiting some stubble on their chins to support their cause—their topic provided them with little ability to talk about wider social issues or to explore multiple perspectives" (Wood & Sheehan, 2020, p. 11).

There is a clear message here that contexts should support nuanced and critical perspective-taking and that they should be significant to the intended learning in some non-trivial way. This is a really good start, but what other characteristics of contexts might need to be kept in mind when complex systems thinking is in scope? I did not find any complexity-framed research with a direct equivalent of Wood and Sheehan's analysis and so I needed to come up with a plan B. That's what I turn to now.

---

[52] At the time of writing a draft histories curriculum in a new format had just been released for consultation. Assuming it is adopted in this format, there will be much stronger guidance about choice of critical contexts.

An exploration of generative types of contexts

My plan B is to work backwards. I will introduce some types of learning contexts that seem to lend themselves to developing complexity thinking. As well as briefly describing selected examples of actual teaching and learning experiences, I'll share some thoughts on the *affordances* of each type of context. What is it about this specific type of context that might open up opportunities to strengthen complex systems thinking?

Contexts that afford a simple practical response to a complex problem

I asked one of my teaching friends to nominate a context that she thought would lend itself to developing complex systems thinking. She teaches home economics, which I think should really be called something like "the sociology of food", and I was confident of an interesting answer. After some thought she instead nominated a context taught by one of her faculty colleagues—an international initiative called Dress a Girl Around the World.[53] In brief, students make a dress for a young girl living in poverty to wear to school. They learn basic sewing techniques. There are no zips or buttons because these could not be repaired or replaced in the context where the dress will be worn. The students select fabric combinations from a provided supply that typically includes "up-cycled" materials (sheeting, duvet covers etc. and donated trims). They also learn about the lives of the girls who are likely to be recipients of the dress they make. Specific attention is paid to why they might not be able to go to school without this specific help, and the difference that having some basic schooling might make to these young girls' lives. A further indication of their difficult circumstances comes from a distinctive label sewn onto the hem of the dress. This is intended to deter people-traffickers by conveying that this is a girl who would be quickly missed.

Intuitively, this context provides interesting possibilities to foster aspects of complexity thinking. For example, what "sensing in" impacts (Chapter 7) might there be from having an unselfish goal for students who mainly live in the midst of plenty, and who are likely to take the right to go to school for granted? My friend said that this aspect certainly caught the attention of the students in this class, prompting considerable

---

53 https://www.dressagirlaroundtheworld.com/

discussion. She noted that the students worked hard to hone their beginner sewing skills to produce attractive well-made dresses. They took great pride in the display of the results. The power of the experience lay in doing something creative for someone else, that is, the opportunity afforded is to develop compassion (Chapter 7). As already sketched, aspects of the Social Studies learning area were also implicated in the context.

In this example the problem being addressed is complex but someone else has decided on the solution. The action to be taken by the students is prescribed and mainly entails making aesthetic choices. I've been thinking about parallels between this example and some common responses to complex environmental issues. For example, in the face of the huge challenges posed by climate change, engaging in actions such as recycling is a common response, even though doing this will actually make a minimal difference. One primary teacher recently described this type of response as a sort of "seduction" (Reynolds, 2020) that was distracting her from addressing deeper and more difficult issues. And yet recycling is obviously a useful thing to do, as is making dresses for young girls living in poverty. So now I'm thinking about this type of prescribed-action context as a sort of "toe in the water". It's great that students can feel empowered by undertaking a positive type of action but what else might need to be considered for a context to offer opportunities for them to build their capabilities for unpacking inherent complexities in an issue, and for taking more personally challenging action?

Contexts that explore open-ended and values-laden issues

> Context-embedded learning involves exploration of socially relevant issues. It supports the achievement of core learning outcomes within national curricula while also enabling young people to engage in, develop understanding of, and act on issues that are impacting their families, schools and communities. (LENScience website: https://www.lenscience.auckland.ac.nz/en/about/teaching-and-learning-resources/Context.html)

LENScience is an education outreach initiative of the Liggins Institute at the University of Auckland. Scientists at Liggins are building knowledge of complex health-related issues, including the various factors that impact on the incidence of non-communicable diseases (NCDs) such as heart disease, diabetes, and obesity. One significant finding from the work at Liggins is that NCDs can have a multigenerational dynamic. The state of health of expectant parents can impact the future health status of unborn children. This happens via complex interactions between the genes and the cellular environment. The biological concept is called epigenetics. At the time of conception the health of the male is implicated, and during pregnancy the female's health status takes on most importance. There is very clear evidence that the mother's own health during pregnancy will impact not only the health of her child, but also any children they might have in generations to come.[54]

I vividly recall a conference session several years ago where I listened to the educators from LENScience talking about their work with adolescents from Auckland schools. Findings of the Liggins research were used as a context for students to undertake their own multifaceted inquiries into NCDs and their potential impacts on their own lives. The LENScience educators reported that one idea in particular really struck home: personal dietary decisions will impact on the health of unborn children over more than one generation. The parents of some students in the pilot also paid attention, changing their food-buying habits out of concern for future grandchildren. Thus the choice of context was an important factor in prompting complex behavioural changes for some families.

On the basis of their experiences in working with adolescents, this group of educators have formulated the following questions about choosing fruitful contexts for student-led inquiries into complex issues:

- Is the context going to be meaningful to adolescents in the target age-group for the particular learning programme?
- Will exploration of the context support development of understanding of and about science, health, and social sciences?
- Does the context relate to an issue for which students of the target age-group potentially have some relevant decision-making power?
- Is the context relevant to the wider community?

---

54 There are many academic papers about epigenetics. If the complexity of this phenomenon and what it means for our personal choices interests you, Vears & D'Abramo (2018), an open-source paper, might be an interesting place to start. Scenario 3 in that paper delves into the considerable complexity of the dietary choices faced by expectant parents.

- Is the context important to the science, health, social science and research communities? Do they want to engage with the public in the communication of research related to this context and do they have evidence to share that will be relevant to the community?[55]

This seems to me to be a useful manifesto for choosing learning contexts that foreground complexity and also resonate with young people's interests and concerns. Now, I want to pick up on the last couple of points to think more about the contribution of contexts that demand community engagement.

Contexts with integral community engagement

Some students who worked with the LENScience materials investigated types of food outlets in their local community. Such investigations typically confirm that many students in lower socioeconomic urban communities are more likely to live in "food deserts". These are places where the readily available outlets mainly sell fast food. A car, or good public transport, is needed to get to places that sell a greater variety of foods, including healthier options. The important point here is that what we eat is not only a matter of making good personal choices. Availability of healthy options is one of a range of factors in the wider community that impact on the food choices people can make (see Vears & D'Abramo, 2018, for other factors). Looked at from a complexity-oriented perspective, NCDs such as obesity are *emergent* health problems (Heke et al., 2019). Students can take some actions, but they cannot solve the complex issue of healthy food provision unilaterally. Some form of community action is implicated.

As well as introducing the idea that complex health problems should be seen as emergent, Chapter 8 conveyed a clear message about the importance of *place-based learning*. Such learning is best supported by the deep practical wisdom of indigenous occupants of a specific place—the concept of "two-eyed seeing" necessarily entails collaboration. Learning in school or ECE settings can provide unique opportunities for students to experience the richness of successful school–community partnerships.

---

55 https://www.lenscience.auckland.ac.nz/en/about/teaching-and-learning-resources/Context.html. Note that a range of potential contexts for school use are listed at the bottom of this web page.

These types of learning experiences are often celebrated in the *Education Gazette Tukutuku Korero*.[56] As a small experiment, I used the article search function on the magazine's website, using just three words: environment, community, and conservation. Here is a brief summary of the first four "hits". There is more to these stories than I can include here. My descriptions focus on the context and the nature of the collaboration:

- As part of the Kids Restore the Kepler (KRTK) programme students at Te Anau primary school monitor traps in nearby native forest, in partnership with the Department of Conservation (DOC) and Fiordland Conservation Trust. The area the students monitor starts 2 kilometres from the school and parents join them on visits. The students check and reload traps, remove rubbish, and monitor signs of bird activity. With students from other local schools they take part in the annual garden bird survey, a citizen-science project (see below). Data from all these activities is systematically collected. The principal is quoted as saying "This partnering activity is very organic and truly grassroots, as it enables all our local schools to use their local environment for learning. Every two or three weeks the students have an interaction with DOC or KRTK, so there is constant involvement with the community". (*Education Gazette*, 2019, *98*(15))

- The Tōtaranui 250 Trust is a voluntary organisation with representatives from the community, in partnership with Ngāti Apa ki te Rā Tō, Ngāti Kuia, Rangitāne o Wairau and Te Āti Awa o Te Waka-a-Māui, and supported by the Marlborough District Council. The trust works with school children in the local area to help them understand the dynamics of first encounters between Captain Cook and local Māori, from the perspective of the indigenous people. Cook and his sailors visited Ship Cove Meretoto multiple times over eight years which meant that encounters between the two cultures were ongoing and complex. Little of this is conveyed in official histories and the Trust and children are working together to create new resources that tell the story of the area from the perspective of the local iwi and hapū. (*Education Gazette*, 2019, *98*(8))

---

56 https://gazette.education.govt.nz/

- For more than 10 years Hurunui College secondary-school students have been helping manage the conservation of native roroa (great spotted kiwi) and whio (blue duck) species in the Nina Valley in Canterbury. In science, students have learnt to use microprocessors to design and trial electronic systems to lure possums to traps. Data collected on chew cards have demonstrated that light and sound lures enhance the smell already provided by the bait in the traps. The additional lures greatly increase the frequency with which pest animals are trapped. Specific projects such as the design of electronic lures are completed in class, but most field work is undertaken voluntarily, once a month in the weekend, with oversight from DOC. Many students stay in the group right through their years at secondary school. The science teacher is quoted as saying "They don't really have strong environmental values at the start, but after being up there and getting the experience, they develop some environmental values and it really changes some of their lives". (*Education Gazette*, 2019, *98*(8))

- "Cultural locatedness" guides Picton Kindergarten's curriculum and is a core part of its teaching and learning philosophy. Many of the children's families are involved in enviro-groups, and they are invited to share their expertise and experiences with the children. One example in this article describes learning about, and looking for, two rowi kiwi recently released into a reserve near the town. A citizen-science project that involved tagging monarch butterflies is another of the learning contexts described. Follow-up studies of the monarch life cycle extended to learning about the menace of introduced wasps. As the head teacher noted, "It's one thing that we're tagging the butterflies but actually the wasps eat the caterpillars, so we're already researching ways of how we can put out traps and catch the wasps and protect the butterflies. We're in the process of making a little butterfly habitat and we've been growing swan plants from seeds." (*Education Gazette*, 2019, *98*(1))

These four snapshots were the first four hits from my rather basic search. Serendipitously, they span all the years of formal education. Children are never too young to engage meaningfully with complex contexts! Although I used the terms *environment* and *conservation*, one of the four stories is a historical-inquiry initiative. Points of commonality across these stories include: learners made repeat visits to local contexts of interest; adults modelled the values and ways of being that they hoped to foster in students; the contexts were important to researchers in the relevant disciplines; and students were supported to experience authentic knowledge-building practices. All of these themes have come up in earlier chapters, as well as being reflected in the Liggins' list of considerations for choosing contexts.

One idea about the importance of collaboration is implicit in the four snapshots, but not specifically discussed. When multiple interested parties collaborate, workable actions to address complex problems are more likely to be conceived and achieved. The e-tools introduced in Chapter 7 were explicitly designed for this purpose. Furthermore, when young people work alongside adults, they learn important knowledge and skills for action-taking that might not come up in other types of contexts. Both the papers cited next are from a recent collection titled "Climate Change and Education for a Sustainable Future" (*Set: Research Information for Teachers*, 2020, (3)).

> In our experience as educators, we have found both adults and young citizens need to learn strategies to effect change with others, while also learning to ask political questions about "how power is exercised in their community, by whom, and with what effect". (Hayward, 2012, cited in Tolbert et al., 2020, p. 56)

> We note that the Climate Change programme overwhelmingly individualises young people's opportunities for taking action. We encourage teachers to supplement this with a focus on opportunities for understanding how democracy works and how children and young people can participate in collectivist strategies, policy debates, and democratic processes. (Eames et al., 2020, p. 45)

Contexts that demand collaborative action are more likely to help students gain insights about the politics and democratic processes involved in addressing complex local issues.

Contexts that introduce the uncertainties inherent in datasets

In the foreword to this book, Markus Luczak-Roesch mentions our mutual involvement in a TLRI research project that explored the use of online citizen-science projects in primary-school classrooms.[57] He said that he chose to co-lead this research, even though he works at the tertiary level, because he intuitively felt this was a context where complex emergence might happen. This comment interested me greatly because I have had that same intuition. And two of the four snapshots above include a citizen-science component, even though I chose the examples rather capriciously. The experiences of the teacher–researchers in the TLRI project suggest another possibility to add to those already discussed.

What are citizen-science projects? What is it about them that might make them a fruitful context for fostering complex systems thinking? In brief, many citizen-science projects are designed by scientists to achieve a goal which they would not otherwise have the person-power to complete.[58] Typically, citizens are involved in the systematic gathering, or interpreting, of data that contributes to a large-scale database. For example Melissa, a teacher in the TLRI research, introduced her Year 5/6 students (10–11 year olds) to the Globe at Night project.[59] This project aims to collect observations of the brightness of the night sky from all over the world, building up a database of light pollution in different places, and raising people's awareness of this issue in the process.[60] Careful measurement protocols are devised to ensure that data are collected under the same conditions as far as possible. Melissa's class soon found out that it is not straightforward to use a brightness chart. Given the same image of the night sky, students initially arrived at different readings of the chart and had to talk about why this might happen and what to do about it. Reflecting on this, Melissa said:

> All the students made independent judgements about the magnitude of the stars visible in the photo. Even though the project provides detailed guides to support judgements, there was a lot of disagreement

---

57 http://www.tlri.org.nz/tlri-research/research-completed/school-sector/citizen-scientists-classroom-investigating-role

58 Sometimes citizens are involved in the design of the project from the start.

59 https://www.globeatnight.org/

60 A collection of other citizen science projects suitable for use in school classrooms can be accessed from the Science Learning Hub: https://www.sciencelearn.org.nz/citizen_science

about which magnitude to place the photos. This was very helpful in supporting students to develop an understanding of the inherent difficulties associated with collecting data via any method that relies on individual judgement. (Anderson et al., 2020, p. 48)

Another teacher (called Teacher A in the paper cited next) also worked with a Year 5/6 class. This class worked on a conservation project. Their challenge was to identify small mammals photographed by motion cameras mounted in tracking tunnels. Like the students in Melissa's class, this group also found out that making accurate judgements was easier said than done. When photographed, animals could be near the camera or further way. If they were moving quickly the image might be blurry. Before the class began to work with the photos a scientist from the project visited to give advice about clues to look for, but students still encountered challenges along the way:

Teacher A noticed students' concern about accurately identifying the images. She highlighted how to record the degree of certainty in the classification on the website, and asked students to consider why it was important to have reliable identifications. She connected providing reliable evidence to "integrity"—a school value with which students were familiar ... participation opened up discussions about data reliability, current limitations of technology, and how gathering data may at times be easy but its interpretation requires careful observation, rational judgement, perseverance, patience, and often involves a degree of uncertainty. (Pierson et al., 2020, p. 23)

Notice the similarities in the challenges reported here. Both these citizen-science projects required young students to *engage with uncertainty* in a way that would never happen during the cut-and-dried "practical work" often found in school-science manuals. Importantly, both teachers actively encouraged discussions about data reliability, and supported students to understand how scientists manage uncertainty in their work.

It seems to me that this sort of experience would complement the uncertainties inherent in exploring complex systems, and hence indirectly help build dispositions for complex systems thinking. Citizen-science projects are not the only ones that could provide such experiences of course. For example, statistical inquiries can be set in rich and meaningful contexts that require students to grapple with similar issues, while also teaching them about how variance in sampling is managed mathematically.

Challenges to consider when choosing a mix of contexts for a programme of learning

Just as I was putting the finishing touches to this chapter, Markus sent me a link to an online discussion of a "social justice" approach to teaching mathematics in classrooms in the United States. The basic idea is outlined as follows:

> As racial inequity soars on the nation's radar, math teachers are increasingly bringing social-justice questions into their classrooms to help students see the subject's relevance and recognize that they can use it to become change agents in the world. Teachers are drawing on high-profile issues such as policing patterns, the spread of the [COVID-19] pandemic, and [election] campaign finance to explore math concepts from place value to proportionality and algebraic functions. They're using math to help students understand phenomena as varied as food deserts, disaster aid, and college-admission test scores (Gerwertz, 2020, p. 14).

The contexts listed are undoubtedly complex and relevant to students' lives. Several of them are politically charged. Gerwertz addresses the potential for opposition, and says there has been some social-media abuse of those advocating for use of this approach. She also outlines *educational* concerns about the potential for a loss of focus on the mathematics itself. There seem to be several threads to this argument:

- that use of these contexts risks glossing over current ill-conceived presentation of maths as a jumble of unrelated ideas
- that a two-tier system will develop where middle-class children learn traditional maths and students of colour learn maths in context
- that the actual maths will get "lost" if learning interest remains focused on the context.

The first two of these bullet points address the traditional curriculum and associated assumptions. But the third is important to the discussion in this chapter. Keeping the intended curriculum content in focus does need to be factored in when choosing complex contexts for learning. Interestingly, one teacher educator cited in the article says that his own inexperience with the approach might have contributed to it not working as well as he anticipated. Changing curriculum (and pedagogy) in the ways described in this chapter is itself complex and entails new learning for teachers. A good place to start might be to ask how the context should *change* the intended curriculum content learning, with a specific type of capability in mind that is a woven mix of both elements. This is essentially the same "weaving" argument as I have used for adding key competencies to the curriculum (see for example Hipkins, 2019b).

A different type of challenge came from Bronwyn Woods when she reviewed this chapter. She noted the predominant emphasis on local contexts in the examples outlined. While endorsing their importance, she also raised the challenge of ensuring that students have opportunities to explore issues at the global level of scale where possible. Local contexts need to be strategically connected to global contexts so that students have opportunities to develop "cosmopolitan citizenship":

> More than ever, we have an intensely connected global community, highly integrated global financial systems, and multinational companies dominating national and international transactions. In environmental politics, human rights, international law and security, and social media, people feel more closely connected than ever before. In this context, students need to understand the multifaceted patterns of economic factors, cultural processes, and social movements that shape their lives. (Osler, 2017, pp. 42–43)

Being clear about the overall learning focus

Planning that includes a clear focus on the intended capabilities to be developed is important to effective use of contexts. But some of the papers I have read caution against defining any type of learning outcome too narrowly. The risk of planning too tightly is that students could be prevented from following non-linear learning trajectories that are personally meaningful for them (e.g., Holbert, 2016). It seems important to be clear about what's in the foreground, and what's in the background (but still important) in any specific unit of work. One benefit of working to develop a clear focus on a specific aspect of systems-thinking capabilities is that students can follow diverse pathways through the complexity of the context, but still all come back to the same central "core" of ideas and/or skills (e.g., chapters 3, 5, and 6). In turn, assessment practices should then also become easier to refocus, regardless of how the students have journeyed though their chosen context. I pick up the assessment challenge in the next chapter.

Chapter 11 reading guide

Strategically chosen learning contexts can open up opportunities to foster awareness, knowledge, and dispositions that strengthen complex systems thinking. Features of such contexts are likely to be some combination of: real-world issues and challenges that have meaning for students' lives and that bring different learning areas together; opportunities for collaboration with members of the wider local community and to take action to address an aspect of a complex issue; and/or opportunities to learn about, and where possible contribute to, complex questions of current researcher interest. The examples included in this chapter show that these types of contexts can be accessible to students of all ages, with appropriate support.

1. Think about a context you have used that proved to be engaging and effective for your intended learning purposes. Which of the features discussed in this chapter did your chosen context include? On a continuum like the one below, where would you place this context in terms of its potential to transform learning?

   Toe-in-the-water ———————————————————— Transformative

2. Did your chosen context *change* the intended learning in a meaningful way? Why or why not? Could you tweak the way you work with this and other contexts to open up further opportunities for complex systems learning?

3. The chapter talks about the "seductiveness" of safe contexts. Contexts that trigger potentially uncomfortable conversations (e.g., about politics, power, race etc.) are often avoided. Is it possible to develop the level of understanding of complexity needed for informed citizenship if contexts that could trigger these types of conversations continue to be avoided? Why or why not? Could ideas about pedagogy introduced in earlier chapters provide a useful starting point for designing learning that, despite being uncomfortable, is able to stretch students' perspectives and contribute to the development of compassionate systems thinking?

Chapter 11: Choosing contexts that support the development of complexity thinking

4. Placed-based learning is exemplified in many of the contexts discussed in this chapter. This makes a strong link back to Chapter 8, which argued that place-based learning is critical to bringing complexity learning and indigenous cultural knowledges together (specifically mātauranga Māori in the New Zealand context). A short (2 pages) research briefing with a focus on teaching about climate change brings both these themes (place-based learning, cultural knowledges) together in the context of learning about a complex issue. You might like to access this briefing, and its three companions, to discuss the issues raised with your colleagues, and to access the suggested resources.[61]

---

61 https://www.nzcer.org.nz/research/publications/connecting-climate-change-place-and-culture-research-briefing-4

Chapter 12
Innovative approaches to assessment

Repurposed assessment approaches can help overcome challenges for capturing meaningful evidence of complex systems thinking. Many different types of learning tasks can potentially contribute to a growing evidence base. An organising framework provides a means of assembling pieces of work as a coherent collection of evidence of learning, and gives students an opportunity to demonstrate awareness of the complexity in play in different contexts.

Some of the curriculum outcomes described in previous chapters could be assessed via traditional assessment tasks. Pencil-and-paper exercises can test conceptual recall and understanding, at least to some degree. Different discipline areas are already likely to have strategies for assessment of students' inquiry activities, and so on. The idea of adapting what is already familiar is attractive, provided certain limitations are kept in mind. Here are some challenges to assessment-as-usual that have already been signalled in the previous chapters.

- Recalling pieces of knowledge does not necessarily constitute evidence of systems thinking. Traditional assessment strategies are like traditional reductionist pedagogies: conceptual understanding is broken down into pieces that are not necessarily reassembled into a dynamic whole.

- Traditional knowledge assessments typically have a "correct" answer in mind. The uncertainties and open-endedness of complex systems dynamics require a different way of thinking about what constitutes an appropriate response.
- Capabilities are demonstrated in action and different capabilities work together as a complex whole. Therefore students need opportunities to show both what they know and what they can do in meaningful contexts.
- The context chosen is likely to impact students' responses.
- The nature of any collaboration taking place is also likely to impact students' responses. Traditional assessment tasks typically look for individual cognitive gains, whereas working together on more open-ended challenges opens up and enables more expansive systems perspectives.
- Capabilities are about *being* as well as doing. However, making inferences about dispositions can be misleading. The learning behaviour that students show will be the result of a complex mix of habits, motivations, interests, and beliefs about the nature of the learning task and so on. Therefore, students need to have an active role in reflecting on their learning and thinking.
- There are clear signals in the earlier chapters that *metacognition* is important, so this could be seen as a legitimate assessment focus if the strategy used allows for thoughtful interactive student input.

I've adapted the suggestions that follow from a range of research papers. Just like teachers, researchers face the challenge of designing data-gathering activities that generate *valid* evidence about demonstrations of specific capabilities. When an aspect of systems thinking is the capability of interest, researchers might bear in mind some of the complexities just outlined, any of which could challenge the validity of their knowledge claims. Unlike teachers, researchers won't necessarily have the advantage of being able to discuss responses with any or all of their research participants. Opportunities for insightful dialogue provide a potential advantage for teachers and I have kept this in mind as I have adapted the creative designs of several research teams to fit better with classroom-based learning/assessment activities.

Assessing gains in knowledge of complex systems

One obvious and familiar source of evidence of learning gains will be that students now *know more* about systems. Such knowledge is important, and therefore worthy of assessment attention. As one research team has noted, complexity concepts constitute the "metacognitive resources required for the development of systems thinking skills" (Gilbert et al., 2019, p. 37). Without a growing knowledge base, students will be hampered in developing the capabilities to be and become systems thinkers in their lives beyond school.

Strategies that help students demonstrate their growing knowledge about complex systems should provide opportunities for them to show part/whole thinking, rather than just recalling separate pieces of knowledge. The ideas that follow are not intended to be exhaustive. I am confident that creative teachers will think of many more ways to adapt and use similar assessment strategies.

Creating systems drawings

Given the prevalence of visual tools in complex systems pedagogy, it is not surprising that visual representations should also feature strongly as assessment strategies. Drawing is an assessment strategy that is accessible to learners of all ages, from the very young (e.g., Curwen et al., 2018), through middle school (e.g., Hmelo-Silver et al., 2017) to adults working together to create more specialised concept maps of complex systems (e.g., Huang et al., 2018).

Drawings done at different stages of the learning can be compared to identify an expanding repertoire of systems concepts. Alternatively, older students might keep adding additional detail to a drawing, making a note of what they have changed, and why, as they go. As just one example, the "cityscape" assessment exercise illustrated in Figure 3 in Chapter 2 could be readily adapted in this way.

Any type of drawing could be done by an individual, or by a group working collaboratively. The latter has the advantage that thinking and assumptions can be shared aloud by the group members. This allows for a metacognitive element to be incorporated into the evidence-gathering process, alongside the growing conceptual knowledge of the group.

Chapter 12: Innovative approaches to assessment

Box 12. An example of drawing as an assessment strategy

One way of scaffolding systems drawings for younger students is to provide a simple outline, to which students add detail. The ARB item shown in Figure 15 is an example of this sort. Several simple questions accompanied the outline, giving students the opportunity to show the thinking behind their additions to the drawing.

This assessment approach was trialled with primary-school students in several different schools. These schools had recently completed a unit on the health of New Zealand's freshwater resources, sponsored by The Royal Society of New Zealand. The students had all spent time exploring a stream near their school and the ARB team was keen to find out what students had noticed and retained from that experience. A detailed analysis of patterns of completion of the drawing can be found on the ARB website, along with four annotated examples of students' drawings.[62]

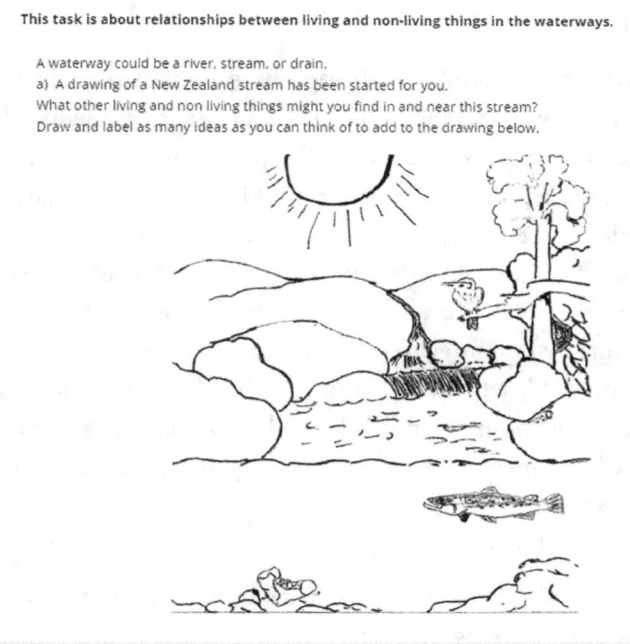

Figure 15. An example of a systems drawing scaffold (Source: ARBS)

---

62 https://arbs.nzcer.org.nz
Note that a log-in is needed to access the ARB item bank, which means direct hyperlink won't work. The item is in the Science bank, and is called *What lives in our waterways?*

Teachers who took part in the ARB Waterways work noted that some students showed more in their drawings than they could tell in written accounts of their learning. Analysis of several hundred drawings found interesting patterns in what students thought to include. Most of them:
- named more animals than plants
- named more animals and plants that lived near waterways than actually in them
- tended to use generic terms (such as fish and insects) rather than name specific species
- named more impacts caused by humans than by natural events.

All ARB items provide this sort of feedback. The specifics are intended to support assessment for learning, by providing clear indicators of possible next steps.

Tackling the part/whole challenge when assessing drawings

What features of drawings will tell teachers that students are actually learning to think in systems terms? Simply adding more detail won't do this so there is no point in creating a marking scheme that says "adds two labels"; "adds three labels", and so forth. Instead, a strategy is needed to notice when the additions represent a significant shift in how students are thinking about systems as complex wholes. One research team addressed this challenge by devising a coding scheme to assess students' before and after drawings (Hmelo-Silver et al., 2017). They used the components, mechanisms, phenomena (CMP) model for teaching students about systems (see Chapter 5) to identify four qualitatively different levels of potential response:

- Draws and names some components.
- Draws and names components and indicates some simple (direct) relationships between them.
- In addition to components and relationships, some mechanisms that drive interactions in the system are named.
- CMP framework is well-established—components, mechanisms, and phenomena are all depicted and linked in annotations.

The questions that accompanied the Waterways drawing task (Figure 15) provided opportunities for students to add detail about relationships and mechanisms, as follows:

> From your picture describe a relationship between: i) two things in the water; ii) one thing on the bank and one thing in the water.
>
> Describe some ways in which human activity can upset relationships in and around this waterway.

In the simplest drawings the researchers saw fragments of knowledge, with few indications that the things drawn were connected via specific relationships. Students who attempted to describe relationships typically drew on their everyday knowledge, but sometimes made errors of fact. It wasn't because they did not have a simple understanding of the concept they were attempting to explain. Rather, their contextual knowledge was weak and thus the mechanisms they described were not accurate (Hipkins et al., 2008). Typically these were feeding mechanisms and they did not know the specifics of local food chains. In this way, the patterns included in their drawings and written answers included indications of their contextual, as well as conceptual, knowledge. The CMP criteria do not specifically allow for assessing both conceptual and contextual knowledge, so this is a challenge to keep in mind.

When we did this ARB work we were not aware of the indigenous knowledge concept of "pattern thought" (see Chapter 8). Looking back, it is not difficult to see how opportunities to repeatedly visit the same waterway, to observe and learn and build pattern thinking both intuitively and with deliberate teaching, would help learners of any age build more robust systems knowledge. At the very least, their contextual knowledge would expand and become more accurate. Interestingly, Huang et al. (2018) specifically used concept mapping to draw out the deep intuitive patterning knowledge held by individuals who were activists for specific local ecosystems that they visited often. Like drawing, concept mapping holds rich possibilities for visually demonstrating learning gains.

Demonstrating awareness of interconnectedness

The following assessment strategies have a focus on identifying multiple connections, or cascading connections. This focus is different from the single event, linear, cause-and-effect thinking that is more typical of traditional assessment activities.

Box 13. Identifying multiple connections, or cascading connections

A number of research teams have devised ways to look for indications that a person can make multiple causal connections as they tell a story about a specific context or scenario. In the ideas that follow I have indicated the age of the participants in the specific research project. Many creative teachers will no doubt see ways to adapt the ideas for different age groups, and for using a strategy as a group exercise rather than only for individuals.

- [Primary] students answered the question: "How could turning on a light in [name of town] eventually affect a polar bear at the North Pole?" The researchers awarded one point for each causal link a student identified (Clark et al., 2017). This process generated a simple score for making connections. The assessment criteria could be adapted along the lines of the CMP criteria above to also assess the systems knowledge being used.

- [Secondary] students wrote an explanation about what might happen if a species of goose was introduced to a local park for the first time ever (Yoon et al., 2019). The researchers looked for instances where students drew on specific complexity concepts in their explanations (see Table 3 in the next chapter). They also judged whether each idea was underpinned by clockwork or complex assumptions about how the world is (see Chapter 4). This was a complex analysis that employed psychometric techniques to bring the data sources together. Bear in mind that it was conducted for research purposes, not assessment purposes. If necessary, an activity like this could be scaffolded for assessment purposes by providing a suggested list of concepts to include, or by providing some starter prompts: e.g., "think about what the geese will eat, and how that might impact the environment, including other species".

- [Tertiary] students wrote an article for a general audience about how "everything is connected" in a specific location (Gilbert et al., 2019). In this study the context was the ocean and the researchers wrote a carefully detailed scenario to which students responded. Early piloting had revealed a need to be specific about the link between the task of making connections and the conceptual content of the module being assessed. With this in mind, the final sentence of the instructions provided to students said "be sure to use systems thinking language and specific examples" (p. 40).

> - [Adults] were given a simple crime scenario. From a list of potential sources of evidence, they indicated those which they thought would be relevant to determining a motive (Petersen et al., 2018). The researchers collated totals to arrive at a broad indication of each person's propensity to see connections. (Note that the provided list of possibilities was very long because the data generated was a basic count of the number of connections made. Students could work with a much shorter list.) Again, I think creative teachers could readily adapt this idea. It might become a group activity where students justified their choices, and perhaps the group then ranked them as more or less likely to add important insights to the investigation. The conversations that ensued would surface causal reasoning, along with any assumptions students might make about direct and indirect connections.

For any of the strategies in Box 13, additional metacognitive insights could be generated via follow-up conversations in which students compare answers. They could gain additional ideas about connections, and also build new insights about the types of connections that can be made. This could be done as a sorting and classifying exercise, once broad categories of connection types had been established (e.g., in the goose example these could include feeding connections, environmental impact connections, competition impact connections, etc.).

Making judgements about "causal extent"

Petersen et al. (2018) used the term "causal extent" to describe the extent to which adults saw causal connections between a simple scenario and several different protagonists. Here is one of the scenarios they gave to their research participants. It was designed for use with adults, but the idea could be readily adapted for students of any age group.

> Oscar has caught a cold and is having very violent sneezing fits during school one day. After one such fit he goes to the bathroom for some tissues. Going into the bathroom he grabs the door knob and covers it with germs he just sneezed into his hands. (Petersen et al., 2018, supplementary information, day 3)

This research team asked their adult participants to indicate to what extent each of the following groups would be affected by Oscar's germs:

the next person that touches the door knob after Oscar; the 5th person that touches the door knob after Oscar; the 10th person that touches the door knob after Oscar; the 100th person that touches the door knob after Oscar; people that touch other door knobs in the school. They also asked their participants to indicate the extent to which they saw all of the following as being responsible for Oscar's germs: Oscar; his parents; Oscar's doctor; the person who designed the bathroom with a door knob. This activity was designed as an online assessment. Participants indicated "perceived impact" by moving a sliding bar along a scale from 0-100 (one sliding bar per item) with a range from "not at all affected" to "very affected". For indications of the "perceived responsibility" of the individuals named, the scale ranged from "not at all responsible" to "very responsible". Overall scores were created for each part by averaging all the responses (Petersen et al., 2018, p. 727).

This strategy suited the team's research purposes, but I suspect it would be more practical for classroom use—and more informative in the moment—if it was adapted somewhat. For example it could be readily adapted for group discussion if each scale was sketched on paper. To aid debate, each suggestion could be written on an individual card, and then the cards could be placed—and moved—on one continuum. Alternatively, for a lively whole-class discussion, students could be asked to take up a place on an opinion continuum that stretched across the room. Either adaptation of the basic activity would surface causal reasoning, making students' thinking visible to themselves and others, and thus aiding metacognitive reflection.

When sharing their thinking about potential causal connections, students might also have opportunities to ask critical questions about scenario details and to display "it depends" thinking (see Chapter 4). For example, the scenario above got me wondering about Oscar's age. I would think about his and his parents' relative culpability quite differently if he was in early primary school (say) than if he was a teenager. In turn, this thought recalled the example of systems thinking using word problems in mathematics, first introduced in Chapter 2 (Salado et al., 2019). *Question asking* about features of the task could be another focus for demonstrating discernment when exploring the causal connections indicated in a specific scenario.

Petersen et al. (2018) used several such scenarios across the multiple days of their adult volunteers' engagement with electronic dashboards showing the dynamics of electricity and water consumption in their community. Based on the "causal extent" measures I have just briefly outlined, they reported that the extent to which people could perceive causal connections could be readily strengthened. They attributed this to the "psychological priming" (p. 733) provided by the dashboards and accompanying resources. However they also found that this exposure did not enhance participants' perceptions of their own embeddedness in other natural systems. They concluded that "causal extent may generalize across domains more readily than one's sense of embeddedness in systems" (p. 733). It seems that seeking and making rich causal connections is necessary, but not sufficient, to achieve the sort of compassionate systems thinking described in Chapter 7.

Designing problem-based assessment tasks

What sorts of assessments might tell us more about students' developing capabilities for compassionate systems thinking? The limitations noted by the designers of the assessment strategy, as just outlined, point to some missing elements. Students would need opportunities to show that they see themselves inside systems, not separate from them. They would need opportunities to show that they can think ethically and act with the sort of concern that seeks an on-balance pathway forward, taking into account conflicting interests and possibilities (including those between humans and the more-than-human world). Even as I type I can see that this could sound impossibly idealistic and certainly unrealistic for younger students. In response to this concern I now introduce a wild card from outside the systems-thinking literature per se. I think the ideas in the paper I am about to introduce describe a potentially practical assessment strategy for students of all ages.

Robert Sternberg is an American educational psychologist. He coined the term "adaptive intelligence" as a specific contrast with the sort of general intelligence typically valued, and measured, within school systems (Sternberg, 2020). He explains that adaptive intelligence allows students to tackle messy problems with unclear solutions. I see that definition as a good way of describing complex issues and so I am drawn to his suggestions for designing tasks that allow students to demonstrate their adaptive intelligence. In common with some of the ideas outlined above, his strategy involves the use of scenarios that students need to work through. Here are two suggested examples, one for students in the American 5th grade, and one for secondary age students, perhaps in a health or science class.

> Tommy and his two close friends in his 5th-grade class have decided that they should listen to teachers and parents only when they want to. What do adults know anyway? They have started purposefully littering, leaving the remains of their lunches scattered around the schoolyard and on adjacent property. They view it as a harmless prank. They don't have to bother finding a litter basket, and they know that, sooner or later, someone will pick up their trash. They are even trying to outdo each other in how much they can get away with. Tommy has been joining his friends, but he is uncomfortable with what they are doing. He is afraid if he says anything to them, they will say he is a sissy and maybe reject him from the group. What should he do?

> A scientific entrepreneur has developed a nutritional supplement that he believes offers great promise for helping people to lose weight. Because it is a supplement to be sold over the counter, it is not subject to FDA (Food and Drug Administration) approval. The entrepreneur has tested the supplement on 200 people, half of whom were randomly assigned to receive the supplement and half of whom received a placebo. Neither group knew which they were receiving. After three months of taking daily pills, the supplement group had lost an average of 21 pounds and the placebo group had lost an average of 1 pound. You have been asked to advise a venture-capital firm that is considering offering the entrepreneur $5.5 million to start producing, marketing, and selling the product. What would you want to ask and/or tell them?" (Sternberg, 2020, n.p.)

Notice the emphasis in the first scenario on balancing one's interests against those of others, and promoting the common good. The second scenario also considers the interests of different groups, and sound ethical practices. But it has an additional focus on question asking, and on using conceptual knowledge gained in the relevant class (e.g., health impacts; regulatory frameworks, and/or procedures used in medical trials).

Sternberg's paper was written for a teacher journal. It includes a table that contrasts the characteristics of problems designed to allow students to show adaptive intelligence with more traditional problems likely to be used in general intelligence tests. I am including this table in full because I think it provides a strong manifesto for designing problems used to assess capabilities for complex systems thinking.

Table 2. Ten key differences between general intelligence and adaptive intelligence problems (Sternberg, 2020, n.p.)

Characteristic	General intelligence	Adaptive intelligence
Type of answer	Right or wrong	More adaptive or less adaptive
Structure	Well structured: A clear pathway to a specific solution	Messy structure: Multiple fuzzy paths to partial solutions
Emotional/ideological Resonance	Low emotional/ideological resonance: Thinking is usually clear	High emotional/ideological resonance: Thinking may be clouded
Life Stakes	Low: Few consequences if a solution is wrong	Often high: A critical need for a good solution
Life Contextualization	Decontextualized problems that are weakly related to life events	Highly contextualized problems that are strongly related to life events
Need for Problem Recognition	None: problems are presented within the test	Great: One has to recognise that the problem even exists
Need for Problem Definition	Low: Problems are usually defined by the test	High: Problems are poorly defined
Time for Solution	Low: Can be solved in a few seconds to a few minutes	High: can be addressed over time but often cannot be definitively solved
Need to Search for Information	Low: Information needed for solution given in the test problem	High: Information needed for solution requires research
Need to Evaluate Information	Low: Information in test problem is generally credible and consistent	High: Information sources are often low in credibility and mutually contradictory

Evidence that students are building systems-thinking habits

Chapter 9 discussed the challenges of building dispositions, and associated habits, to be and become complex systems thinkers. Could this be an assessment focus in its own right? I think we need to tread very carefully here. As I noted in the introduction to this chapter, making inferences about dispositions can be misleading. The learning behaviour that students show will be the result of a complex mix of habits, motivations, interests, beliefs about the nature of the learning task and so on. Only students can explain why they have responded to an assessment task in a specific manner, which implies that they need to have an active role in reflecting on their learning and thinking.

Students do need guidance about what to look for as they reflect on their growing systems-thinking capabilities. Such guidance could focus on the sort of evidence that will show students where and how they are developing the habits of a systems thinker—discerning complexity in contexts where others might see (or simply assume) more linear causality at work. The habits of systems thinkers introduced in Chapter 9 could provide one basis for such guidance. Here is another model that might be useful. Petersen et al. (2018) drew on a systematic review of books about systems thinking to identify six "orientations" that suggest an inclination to think in terms of systems connections. Their six orientations are:

Causality: a person who is oriented to systems thinking is likely to identify multiple causes for things that happen within a system.

Logic: a person who is oriented to systems thinking is likely to expect and seek logical explanations for how interactions within a system may impact the behaviour of the whole system.

Data: a person who is oriented to systems thinking is likely to look for data associated with events, structures, and patterns within systems.

Structure: a person who is oriented to systems thinking can identify explicit and implicit systems structures.

Subjectivity: a person who is oriented to systems thinking understands that our subjective mental models of how systems work influence our observations, conclusions, and individual behaviour.

Self-reflection: a person who is oriented to systems thinking understands that we are embedded in, influenced by, and have effects on systems we interact with.

These criteria describe adult orientations, but students could notice and document instances when they use systems-thinking strategies and tools in new ways or in new contexts, as appropriate to their age level. For example the ladder of inference tool introduced in Chapter 5 and the iceberg tool introduced in Chapter 9 would both provide opportunities for students to show awareness of their own and other's subjectivity, and to show that they can reflect on the implications of this expanding awareness.

Evidence of a strengthened disposition for complex systems thinking will grow gradually, across multiple learning contexts and tasks. This thought, in turn, suggests that students will need to be supported to organise and keep track of their learning gains over longer time periods. Building a portfolio of evidence is likely to be the most practical way to do this.

Organising a portfolio of evidence

Pioneering teachers in the compassionate systems thinking initiative have been trialling innovative approaches to assessment of complex systems thinking. Student portfolios are the tool they are working with. Use of portfolios is well established in many schools so that is not necessarily innovative per se. What is different is the organising framework they are exploring. With their feedback and critique, this framework has undergone several iterations and currently uses the acronym ACT.[63]

Add: new knowledge and skills

Connect: conceptually, emotionally, systematically, in ways that build understanding

Transfer: build a deeper understanding by transferring to new contexts—developing discernment in action.

The ACT framework has been designed to support both students and teachers to notice and appreciate learning gains in any subject area. It provides a structure for gathering and curating artefacts that capture personally significant learning experiences. Students can catch those moments when they made a personally significant link between a new idea and what they already knew, or when they transferred an important new insight to a different context. Perhaps they realised that a concept or

---

63 Jane Drake has generously given me permission to use this framework, which remains a work in progress and is yet to be formally published.

skill they had previously learnt could be usefully redeployed to meet the demands of a new learning challenge.

The idea of "discernment in action" captures this sense that students are aware of important learning gains in what they know and can do—they can select the most appropriate knowledge to use to get the outcome(s) that they envisage. Another aspect of discernment relates to making choices about what to include in a portfolio. Including an overwhelming amount of material will not be helpful for anyone and so the focus should be on personally significant examples. If all this seems like a big ask for younger students, one practical suggestion is that teachers provide options that students can choose between, and then justify those choices.[64]

An element of metacognition is implied in the selection and annotation of work to be included in a portfolio, but the main focus of reflection is on how new knowledge is being put to work. Metacognitive awareness is particularly important in complexity contexts because "everyday" ways of thinking might need to be consciously set aside, and starting assumptions will impact the way that more complex causal reasoning unfolds. Heinrich and Kupers (2019) used the term "knock-on effects" to describe the sort of evidence that is called *transfer* in more formal theoretical terms. Basically such evidence would show that students can use their systems-thinking capabilities in different contexts, or in new ways in familiar contexts. The sorts of evidence that Heinrich and Kupers sought are repeated here. Students might show they are:

- actively looking for interconnections when solving problems
- displaying a higher sensitivity for systems as wholes
- asking different types of questions
- valuing discussion of complexity models in subject lessons
- finding connections between learning contexts and daily events
- critically engaging their own assumptions and worldviews
- gaining new perspectives on societal challenges

---

64 My thanks to Jane for this suggestion.

This list of actions could be adapted as a set of prompts for students when they are reflecting on how they are using their growing systems knowledge, in order to add to their portfolio.

Being clear about assessment purposes

Reading back over this chapter, I have become aware of my own assumption that a portfolio might be assembled, first and foremost, to support *assessment for learning*. Suggestions for collaborative conversations are formative in intent. The different perspectives that are shared allow all students to expand their thinking about the systems dynamics of the context being discussed. The use of prompts to reflection provides support for students as they choose appropriate pieces of work to include in their portfolio, and guides their reflective writing. And so on. But what happens if it becomes necessary to make a summative judgement for reporting purposes? Perhaps the school's board of trustees requires a summary statement about the progress being made by students, and therefore a judgement must be made about how much progress each of them has made. How might the evidence collected in a portfolio be used to make such a judgement, especially since we must assume that the whole will be more than the sum of the parts? This is the challenge I turn to in the next chapter.

Chapter 12 reading guide

Repurposed assessment approaches can help overcome challenges for capturing meaningful evidence of complex systems thinking. Many different types of learning tasks can potentially contribute to a growing evidence base. An organising framework provides a means of assembling pieces of work as a coherent collection of evidence of learning, and gives students an opportunity to demonstrate awareness of the complexity in play in different contexts.

1. This chapter begins with a discussion of the limitations of traditional knowledge assessments, arguing that they cannot meaningful capture growth in complex systems thinking. Do the reasons given resonate with your personal experience of assessment of other competencies/capabilities? Are some of these points more convincing to you than others? If so, why might that be?

2. Many of the less traditional tasks described in this chapter can include opportunities for group discussion, so that thinking can be more visible. This is particularly important when thinking itself is the assessment focus. What is your experience of working with tasks of the sort described? Which strategies appeal and which don't? What practical advice would you give a beginning teacher about working with interactive assessment strategies?

3. Table 2 outlines features of problem-solving tasks that are designed to give students opportunities to demonstrate their "adaptive intelligence". In what ways do these features reinforce key pedagogical messages from earlier chapters? Could you begin to build a bank of effective problem-solving tasks that will work for students in your context?

4. Is the idea of an organising framework for student portfolios useful? Why or why not? What are the pros and cons of building a portfolio of evidence to demonstrate growth in complex systems thinking?

Chapter 13

Indications that students are making progress

Progression is a complex phenomenon. The learning progress made by individual students is seldom linear and often unpredictable. Nevertheless, there are research-informed insights about important developmental step-ups in complex systems thinking. A part/whole approach to bringing these insights together suggests a practical way of capturing evidence of overall progress.

When I first drafted the assessment chapter, I tacked a section about progression onto the end. I had found bits and pieces of information, each of which seemed useful in its own way. However the realisation gradually dawned that I needed to treat this topic as a part/whole challenge if I was going to be consistent with the message that part/whole thinking is important, which is threaded throughout the book. Sauce for the goose is sauce for the gander! That little "aha" moment resulted in the splitting out of this rather more speculative chapter about progression from the assessment chapter. I begin by outlining several important ideas about the complex nature of progression. I then relate these warnings about complexity to the various models of progression in complex systems thinking that I have found in the literature. The final part of the chapter makes some suggestions about what teachers could do to capture the progress students make over time, regardless of a student's chronological age when starting out on the journey to complex systems thinking.

Progression is a complex phenomenon

> Learning progressions are ... conjectural or hypothetical model pathways of learning over periods of time that have been empirically validated. (Duschl et al., 2011, p. 124)

This definition of learning progressions is slightly abbreviated from the original. It captures the idea that a progression is a *model*—it maps out potential ways in which students' understandings and abilities might change and build over time. To be "empirically validated", data captured from actual students' progress need to fit the model and need to be seen to be soundly based. This is sensible advice but when it comes to a new learning focus—like complex systems thinking—there is a "chicken-and-egg" dilemma in play. You can't catch evidence of what students can realistically be expected to do if the thing of interest is something they have never been asked to do before! As one illustration, Chapter 11 provides an example of how primary-school children taking part in citizen-science projects were able to critique evidence in ways that were well beyond typical expectations for learners of their age. (They were given this opportunity in an accessible context, with an appropriate level of teacher support.) This chicken-and-egg challenge is just the first of several dilemmas that need to be considered when building models of progression.

As we saw in Chapter 7, there is always an element of conjecture to models. They are guides to action rather than "real things" in the world. Certain assumptions about how the world itself "is" are built into any models we construct and use. In the case of learning progressions, potential models need to take account of some important ideas about learning, assessment, and progress. First, learning itself is complex and non-linear. The next chapter explains why children's thinking can develop in a wide range of ways—not always as the teacher intends. Ideally, students' *actual learning trajectories* need to be considered alongside any logical, sequenced models of progress. It is simply not possible to capture every potential learning pathway for every child, but where common development patterns can be detected progressions could help teachers anticipate and plan for conceptual difficulties that might arise.

A second and related point is that progressions should support teachers' use of assessment for learning practices (Duschl, 2019).[65] To be practically useful, learning progressions need to provide support for teachers' and students' noticing and decision making. This implies that supporting metacognition for example (see Chapter 5) would require a progression that identifies visible thinking patterns which signal when students are ready to be challenged with next steps in complex systems thinking. Discussion of their current thinking patterns would support students to reflect on their achievements thus far and what they might need to focus on next. Several recent critical analyses of the whole field of progressions research have described this approach as a "work with it" way of modelling progress (Duschl et al., 2011).[66] I have endeavoured to keep this "work with it" sensibility in mind as I documented the progressions research outlined below.

Here's the third important point to consider. Earlier chapters have drawn attention to the importance of developing capabilities for complex systems thinking. These capabilities are understood to be *more than* higher levels of content knowledge—skills and dispositions are also implicated. Some researchers believe that the development of more complex capabilities should be a focus for progressions in their own right. Early work in this area has tended to focus on several aspects of critical-thinking capabilities: argumentation; thinking with models; and/or epistemic reasoning (i.e., thinking about how we know what we know). As I will outline shortly, there is some speculative work with a specific focus on capabilities for complex systems thinking, and it draws on these types of critical-thinking sources.

As if that's not enough complexity to deal with, sociocultural learning theory says that we need to consider how each child's learning is supported in the context in which they do that learning. Assessment

---

[65] The converse should also hold—when teachers use robust assessment-for-learning pedagogies, the insights they gain will help to inform their judgements about whether and how students are making progress. My thanks to my colleague Mohammed Alansari for spotting this point.

[66] The contrast implied is with a "fix it" model of progress where students are judged against a set sequence in which ideas or skills should develop.

needs to make space for students to show what they can *actually do* in meaningful contexts and to show they can transfer their ideas and skills to less familiar contexts. One Australian researcher argues that progress becomes visible when students use their new learning in "cultural landscapes" beyond the classroom (Tytler, 2018). In this way of framing the progression challenge, descriptors of progress need to take account of the openness and authenticity of the context (see Chapter 11), alongside how the student responds and adapts.

Different models of progression (and how they might fit together)

Across the curriculum in general, the increasing complexity of a specific feature (idea, context, skill, etc.) is often used as a criterion in established models of progression. This is a dilemma when complexity itself is the focus. We can hardly talk about differentiating between "simple complexity" and "complex complexity". Instead, the models I have found consider how various pieces of evidence might come together to describe progress in developing complex systems thinking as a phenomenon in its own right. I now outline four different ways of framing this challenge, beginning with research that focuses on concepts that are used to describe complex systems.

As you will shortly see, these models all come from science-education researchers. This probably reflects the inclusion of "systems and systems models" as a cross-cutting concept in the Common Core Standards for science in the United States.[67] One short conference paper I found promised to be about assessing progression "across disciplines", but the three simulations discussed were all set in science contexts, albeit in different science disciplines (Rehmat et al., 2020). Complex systems concepts do apply to both physical and social systems, so I am hopeful that the progressions outlined below might be broadly applicable across different subject contexts. This is a conversation that I hope social scientists and others will take up.

---

67 See https://www.nap.edu/read/18290/chapter/13. Directly or indirectly, most of the other cross-cutting concepts are also implicated in systems thinking: patterns; cause and effect; scale, proportion and quantity; energy and matter—flows, cycles and conservation; structure and function; stability and change.

A developmental sequence in understanding and applying concepts

The idea of making progress implies *developmental* change. An aspect of the way in which students use their growing complexity knowledge becomes qualitatively more advanced. In turn, this advance is considered to be important because it takes the learner closer to the ultimate goal or purpose for developing complex systems thinking. To illustrate, Table 1 in Chapter 4 contrasts aspects of linear/clockwork thinking with the equivalent non-linear/complexity thinking (Yoon, 2008). It is a very big developmental step-up from mainly thinking in linear/clockwork terms to mainly thinking in terms of dynamic complexity. Since being able to do the latter is important to informed citizenship when dealing with complex issues, this is an important developmental milestone to consider.

Recently, Yoon and her colleagues have used this contrast between a clockwork and a complexity orientation to look for evidence of a developmental progression in students' likely understanding of concepts that explain the *dynamic behaviour* of complex systems (Yoon et al., 2019). They identified six concepts that they anticipated might need to develop *sequentially*. These concepts are: scaling effects; networked interactions; multiple causes; dynamic processes; order; and deterministic effects. They worked with 44 student volunteers, who spanned Grades 8–12 in the United States' school system (broadly, secondary-school levels in the New Zealand system). Each student wrote an explanation about what might happen if a species of goose was introduced to a local park for the first time ever. The use of a familiar local context was deliberate, so that students could display their knowledge of systems concepts in context, and in combination with one another. A multiple-step process of psychometric analysis was used to arrive at the proposed model of progression shown in Table 3. The examples below each row are taken verbatim from the research paper (pp. 12–13).

Table 3. A proposed developmental sequence for understanding concepts that explain systems dynamics (after Yoon et al., 2019)

Concepts in developmental order	Students show evidence of understanding that:
Scaling effects	Small changes can lead to large effects, including cascading effects; other variables in the system must be considered
The geese may chase off other animals, which could stop them from eating plants they normally eat, which causes the population of these plants to increase. An increase of other animals that feed on these plants may then occur as they have more to eat. [cascading effect]. There may be overpopulation of the geese as they lack the natural predators [large scale].	
Networked interaction	Non-linear emergent effects come about as a consequence of interdependencies in the system
The geese will probably help the ecosystem. First, their droppings might make the soil more fertile, and plants will grow better [nonlinear]. There may be more O_2 as a result [interdependency]. The result of O_2 and plant increase could cause a wet and warm ecosystem [emergent patterns].	
Multiple causes	One change can trigger multiple others, and they then cause more, which leads to emergence
The geese may also eat most of the grass [cause]. The caterpillars and other grass eaters will die or move to another ecosystem [effect/cause]. This would mean that the decomposers will have less to eat [effect/cause]. The soil may have fewer nutrients as a result, and the trees will grow less well [cause/effect]. The geese may also damage statues and walkways with their droppings [cause/effect].	
Dynamic processes	Change does not happen once and then stop, there are ongoing changes as the system responds and adjusts
The geese may increase the competition for the same food with other animals. The other animals may leave the park to seek greener pastures. They and the geese may also simply starve, and their populations decrease. However, over time, with more geese in the park, the amount of nutrients in the soil is likely to increase as there are more decaying matter (faeces and dead geese). This allows the park to support more producers and consumers. At the same time, overcrowding may occur. The lack of space may again decrease the populations. This process will continue on until a new equilibrium can be reached or the cycle can carry on indefinitely [continual changes].	
Order [within the overall system]	Even if one species is central to the discussion, decentralised organisation means that everything in the system has to be taken into account
Geese may have both positive and negative effects on the ecosystem. For example, they [a central actor] may decrease the amount of food available to other animals who preyed on the same type of food as the geese. The geese may also increase the amount of food available to animals who can eat the geese. The effects of geese on the ecosystem cannot be easily determined without considering the other animals and plants in the same ecosystem as well [while only one actor is explained, it shows evidence of decentralisation in understanding].	

Concepts in developmental order	Students show evidence of understanding that:
Deterministic effects	It is impossible to say exactly how a system will be impacted if the change is one that has never happened before. Uncertainty and unpredictability are acknowledged and emergent change will only become apparent over time
Since the geese arrive at a place they haven't ever been before, there are many ways they can affect the ecosystem and it is impossible to say exactly how [uncertainty in tone]. For example, they could drive other birds away so that they can lay eggs [1st alternative]. They could drive other birds away because they compete for the same kind of food [2nd alternative]. They could cause the increase of other animals [3rd alternative] who feed on geese. They could cause the increase of other birds [4th alternative] because the geese have become an alternative food source for existing predators. It's really hard to tell [suggestions of complete unpredictability].	

Yoon and her colleagues said they regarded this sequence as provisional until they had a chance to test it in actual classrooms. However they noted that it broadly accords with other recently published studies such as the CMP framework outlined in chapters 5 and 12 (Hmelo-Silver et al., 2017).

In the theoretical language of progressions, their model could be described as having a reasonably fine "grain size". There is quite a lot of detail to guide "next steps" in teaching and learning. Specifically, the model serves as a reminder to check that more foundational ideas such as scaling effects are in place before students tackle harder ideas such as deterministic effects. In reality, individual students will seldom demonstrate "pure" sequential progress in assimilating these ideas. Sometimes they will think to apply a mix of ideas appropriately and at other times they might not recognise the opportunity. Learning is messy and an on-balance judgement is needed if this sort of model is used to report on students' progress. The next model at least partially tackles the grain size challenge, by discussing use of concepts in combination with one another.

Using concepts in combination as an indicator of progression

Chapter 3 outlined complexity concepts and indicated that some are more difficult to grasp than others. Samon and Levy (2020) focused on four specific systems dynamics that are known to be more difficult to understand: emergence; stochastic (random) behaviours; multiple levels of organisation; and decentralised control. They undertook a detailed analysis of relationships between these concepts, as depicted in students' drawings of diffusion. They proposed three levels of sophistication in understanding, based on the way in which students combined these concepts into coherent mental models:

1. **Single level thinking**: mainly non-sophisticated reasoning about all four concepts.

2. **Dual-level thinking:** non-sophisticated reasoning about emergence and stochastic behaviours, combined with more sophisticated reasoning about levels and decentralised control. Samon and Levy identify the tendency to think anthropomorphically as an issue here—for example students might say particles "want" to go where there are fewer of them. Yet these students have often understood that particles do this individually and without anything or anyone being in charge, so they do understand decentralised control. Their next challenge is to understand random behaviour and emergence.

3. **Emergent thinking**: more sophisticated reasoning about all four concepts. Typically, only older secondary students would show this level of coherence.

Notice that the term *mental models* is used in this suggested progression. This implies that making progress is about how ideas are used in a *stable combination*—how students have personally organised them into a coherent framework of ideas that we might label as "understanding". I'll come back to the complexities of cognition in the next chapter, including the idea of how mental models are formed and can be modified.

Notice also that Samon and Levy's proposed sequence is set in the context of a system where the focus was just one main type of variable (particles that move) whereas an open-ended number of variables could be drawn into the discussion about introduction of geese to a park. This is another reminder that the task context will have an impact on students' ability to display their conceptual understanding of how systems behave. The next set of ideas draws contexts into a proposed developmental sequence in combination with concepts, creating a somewhat bigger grain size again.

Adding contextual knowledge to the components, mechanisms, phenomena (CMP) model

In the CMP model there are four qualitatively distinct levels at which students can show knowledge of complex systems (see Chapter 12). Both Chapter 11 and Chapter 12 also discuss the important role that the choice of learning context plays in creating opportunities for students to explain their ideas about a system. Can we bring concepts and contexts together in one coherent model? In Table 4 I have attempted a speculative synthesis, based on two quite different research programmes.

Table 4 juxtaposes the CMP framework with research that identifies three levels at which students' might assimilate their contextual experiences into their mental models of concepts (Lemmer et al., 2020). Lemmer et al. propose three developmental steps when integrating contextual / experiential knowledge with ideas. In Table 4 I have speculated that these might line up with increasing sophistication of understanding of a specific system. The left-hand column is my attempt to catch the essence of each step up in this progression sequence.

Table 4. Potential indicators that students have made progress in building their conceptual and contextual knowledge of a complex system

Learning behaviours (my synthesis)	Features of a drawing (after Hmelo-Silver et al., 2017, CMP model)	Features of students' mental models (after Lemmer et al., 2020)
Showing awareness of some system components	Draws and names some familiar components (C)	Intuitive fragments of knowledge arise from everyday experiences and are weakly structured
Building a growing library of experiences; identifying more connections between components of a system	Draws and names components and indicates some simple (direct) relationships between them	Students attempt to integrate systems concepts with their personal/ experiential knowledge base
Describing impact of interconnections, drawing on new conceptual and contextual knowledge	In addition to components and relationships, some mechanisms (M) that drive interactions in the system are named	
A coherent dynamic system can be clearly described and modified to include new ideas	CMP framework is well established—components, mechanisms and phenomena (P) are all depicted and linked in annotations	Mental models are well established and used reasonably consistently in explanations

My synthesis is speculative but I think it makes intuitive sense. It aligns with work our wider team has undertaken that focuses on the importance of building a "library of experiences" as an important foundation for later more abstract learning (Bull, 2011). These broader signposts could provide useful indications of how school-age students make progress in building more comprehensive and coherent mental models of complex systems, based on both their conceptual understanding and their contextual experiences and knowledge. At this coarser grain size, fluctuations in the use of specific ideas are not as visible, but overall judgements might be easier to make.

Progression based on what students can do with their learning

The models just outlined all have a main focus on conceptual progress. This is important but it should not be the whole of the progression story. What students can do with their growing knowledge base also matters. This thought directs attention to the *purposes* that are considered important when arguing for the inclusion of complex systems thinking in the overall curriculum. A clear and consistent theme throughout this book is that those purposes relate to informed citizenship—being able to recognise complexity at work in the world and take it into account when making, and acting on, personal decisions and taking collective action on matters of concern to a community. The model I now introduce does take direct account of citizenship purposes. It is based on the idea of "systems literacy" [for acting in the world] (Plate & Monroe, 2014).

Plate and Munroe describe seven "skills" that make up systems literacy. They say that these skills are needed for informed citizenship and they are also important for scientists and other researchers and policy makers who work with complexity. I have paraphrased their descriptions here. Note that I put the word "skills" in quotes because arguably the first five are essentially understandings and dispositions. The action to be taken is implied rather than explicit.

- **Recognising interconnections:** not having tunnel vision, or focusing on narrow chains of causality, or failing to apprehend impacts of personal actions.
- **Identifying feedback:** understanding how feedback loops impact on the behaviour of a complex system.

- **Understanding systems at different levels of scale:** being able to zoom out and understand a system at the broad scale, then zooming back into the details (i.e., working between parts and wholes).
- **Differentiating types of stock and flows:** complex systems often show delayed responses, which can be more readily understood when a system is seen as a series of stocks and flows.
- **Understanding dynamic behaviour:** systems rhythms often include long periods of stability, punctuated by rapid changes. If our mental models are static, these rhythms are unlikely to be factored into decision making.
- **Creating simulation models:** working with simulation models is important, but making and testing models is even more demanding.
- **Incorporating systems thinking into policies:** skills 1–5 strengthen decision-making processes, which now need to be applied to making informed decisions about complex systems in real life.

Plate and Munroe drew on research in the related areas of critical thinking and higher order thinking to build progressions for each of these seven areas. They argue that each of them is sufficiently distinct to require separate assessment, which is why they provide seven separate progressions. There is a certain practical appeal to this argument. Examples in each category are provided throughout this book and specific assessment suggestions for most of them are included in Chapter 12. But treating each one separately, while practical in one way, cannot address the challenge of how they come together as a unified action-taking whole. With this challenge in mind, I looked for a pattern across the seven separate progressions—could I see the "essence" of the action-taking implied at each level? Table 5 is my attempt to characterise this.

Table 5. A synthesis of ideas about skills progression (after Plate and Munroe, 2014)

Name given to stage	How "skill" is manifested (my summary)
Below basic systems literacy	Draws on linear, static, models with little or no understanding of complexity, or ability to factor it in when making decisions.
Basic systems literacy	Has a basic conceptual understanding of skills 1–5, can interpret and use simple models and apply them to personal decision making.
Intermediate systems literacy	Understands the dynamics of systems across a range of variables in multiple contexts, and across various models and tools, including those with quantitative features. Factors such dynamics into personal decision making and understanding of policies.
Advanced systems literacy	Can hypothesise, build, and test complex quantitative models of systems. Uses these models as tools to evaluate and decide between competing policies.

The "advanced" level on this progression describes the sorts of actions needed by scientists and policy makers (see also Chapter 15). The "intermediate" level is arguably needed for informed citizenship and hence would be a useful goal for the senior secondary school level. In the language of progressions, this would be the "upper anchor" for any model of progression across all the years of school. Big conceptual milestones on the way to this goal are described in the models introduced above. What this model adds is a focus on what students can do with what they know.

A practical strategy for noticing and documenting progress

As I was wrangling this chapter into shape, I happened to attend a workshop that explored ways to capture students' progress in meeting the "progress outcomes" of the digital technologies curriculum.[68] The context was a meeting of the teacher/researcher team involved in a TLRI project exploring the use of online citizen-science initiatives in the classroom. This second project in a series has a focus on opportunities to achieve digital outcomes as well as building science capabilities for citizenship.[69] The practical challenge for the teachers in the team is very real. How can

---

68 https://nzcurriculum.tki.org.nz/The-New-Zealand-Curriculum/Technology/Progress-outcomes

69 http://www.tlri.org.nz/tlri-research/research-progress/school-sector/on2science-multiple-affordances-learning-through

they provide robust evidence of gains in students' computational thinking and/or digital fluency (the two areas of progress in focus) as they contribute to online citizen-science projects?

Stephen Ross, the adviser leading the workshop, carefully unpacked the milestones suggested in each progression. He then introduced short, workable lists of learning behaviours teachers might expect to see by each milestone. He suggested that the teachers could use a simple Google Form to quickly catch instances when students showed these behaviours in class. An advantage of doing this is that students can show learning related to different milestone points at the same time. This addresses the challenge that their learning won't come neatly prepackaged in discrete "stages". Entries on teachers' forms could then populate a web-based spreadsheet that collects overall data. The *pattern* of learning gains that accumulate over time is what matters, and a spreadsheet can catch these without taking up additional time. Another advantage is that the pattern of *opportunities* is also revealed. If one aspect of a progression has been caught multiple times but another possibility has been overlooked this should also show up on the spreadsheet.

I've outlined this idea at some length because I think it could work for capturing progress in complex systems thinking. However, the chicken-and-egg dilemma kicks in at this point. We don't have access to a ready-made progression that has been empirically validated. (By contrast, the unequal spacing of the milestones in the digital-technologies progressions is one indicator that they have been subjected to psychometric testing and confirmation.) Furthermore, students could be at different stages of their overall schooling before they are introduced to complex systems thinking. I was an adult before I had this opportunity. Presumably, more mature students might be expected to make faster progress than younger ones, and to grasp sophisticated concepts more readily. In the meantime, the models outlined in this chapter could be used, in combination with specific learning plans, to create lists of potential learning outcomes that teachers might support their students to develop and demonstrate.

There still remains the issue of integrating insights from the various models, with their different foci. In the absence of a systematic programme of research to empirically validate a full range of progression milestones, the best I can do for now is to draw on yet another research insight to suggest a rough sequence and likely overlaps. Sommer and Lucken (2010) make an important set of observations about the comparative ease with which students aged from 9 to 11 can model the organisation of systems compared to working with system properties such as emergence and dynamic effects. As you might anticipate, the former is easier for students of this age than the latter. Working with this insight, I have constructed a tentative model (Figure 16) that broadly sequences the models of progression described above.

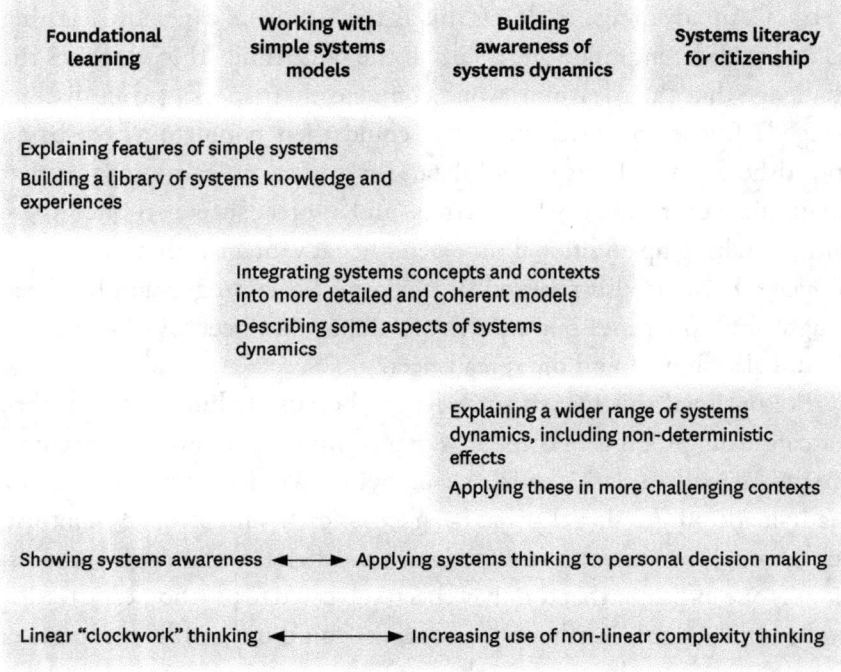

Figure 16. A schematic overview of progression in complex systems thinking

The complexities of meaning-making

For me, the realisation that researchers were treating the question of progression as a complex phenomenon in its own right came as a bit of a "lightbulb" moment. To be honest, it was a relief. I have long believed that teachers can't be expected to hold many different models of progression in their heads while working—that what is needed is a sort of a roadmap with a few important signposts to guide the unfolding learning action. I think (I hope) this is what research teams investigating progression in complex systems thinking are moving towards.

There are several important ideas about meaning-making in the short paragraph immediately above. What we call lightbulb or "aha" moments are times when we abruptly reorganise our mental models to understand the same ideas and events in a new way. Considerations about how much detail we can "hold in our heads" are explained by the limitations of our working or short-term memory. These and many other related ideas about our meaning-making (or cognition) are part and parcel of the metacognitive resources that teachers bring to their work. In the next chapter I discuss cognition as a complex phenomenon, and briefly explore a range of ways in which this knowledge can be purposefully deployed to help students overcome known learning challenges such as those outlined throughout this book.

Chapter 13 reading guide

Progression is a complex phenomenon. The learning progress made by individual students is seldom linear and often unpredictable. Nevertheless, there are research-informed insights about important developmental step-ups in complex systems thinking. A part/whole approach to bringing these insights together suggests a practical way of capturing evidence of overall progress.

At the time of working on this reading guide, New Zealand's "curriculum refresh" process was just getting underway. The content for two prototype subjects (digital technologies, and Aotearoa New Zealand histories) had already been structured as broad sets of progressions. These prototypes represent a significant departure from the outcomes-based structures of the New Zealand national curriculum framework, which have been in place for the past 20+ years. I was part of a team who wrote a background paper about making the change to a progressions approach across the whole curriculum, and what it is intended to achieve.[70] If you have not already done so, it could be helpful to access and read this paper alongside your thinking about Chapter 13.

1. What do you think the term *making progress* means? Why is making progress a deceptively simple idea? What makes progress tricky to document from students' demonstrations of learning? What are the benefits and drawbacks of assessing achievement using a progressions approach?

---

70 Chamberlain, M.; Darr, C., Hipkins, R., McKinley, S., Murphy, H. & Sinnema, C. (2021). *New Zealand Curriculum Refresh: Progressions approach*. Ministry of Education.

2. Figure 16 provides a broad schematic overview of progression in complex systems thinking. Does this speculative model make sense to you? If you were going to modify it in any way, what would you change and why? Where would you have positioned your own understanding of making progress in complex systems thinking at the beginning of reading this book compared to now?[71]

3. How well do you think this model would work with the portfolio framework introduced in Chapter 12? What extra detail might be needed in order to make the model really useful?

4. Have you had a chance to try out some of the teaching and learning ideas from earlier sections? If yes, can you annotate the model with examples of actual learning behaviours you have seen? Where, for example, would you position basic demonstrations of "it depends" thinking? (see Chapter 4).

---

71 Thinking through this question has the potential to highlight pieces that could be important but are missing in the progressions research I have drawn on. For example, if I was to place myself at the start of the writing process and again now, I would want to show a significant shift in my understanding of what it means to see myself inside systems, and decentered within them at that (see chapters 7 and 8). I did not find any research that explored that specific progress shift.

Chapter 14
Cognition is complex

This chapter explores *learning* through a complexity lens. Understanding cognition as a complex phenomenon can be helpful for finding strategies to support students when they face a range of learning challenges—or when they need to be challenged and extended. Relevant insights range from the micro-level (neurobiology) to the macro-level (ecologies of learning).

This chapter has a focus on acts of learning, aka cognition itself. This is yet another huge topic that could be a whole book in its own right. When I was a science and biology teacher, I could never understand why so little attention was given to the biology of learning. It always seemed to me that this was important knowledge that every student should be able to access—and I only had in mind classical theory about how the brain works. The need is even more urgent with recent developments in the field of neurobiology, and so I begin this chapter with a brief outline of contemporary thinking about the brain as a complex adaptive system. Some implications for teaching and learning are then outlined.

The psychology of learning is now understood to be much more complex than people used to think. Chapter 9 has already provided some indications of recent shifts in understanding: in that chapter I outlined changes in the ways that habits and dispositions are understood. In this chapter I introduce research that starts with the assumption that cognition is complex and explores more effective ways to teach topics that are known to be difficult for students to grasp. The chapter concludes with a focus on cutting-edge research on how to shift students' lived understanding of complexity at play in their lives.

Understanding the brain as a complex adaptive system

In the 1980s a research scientist called Gerald Edelman proposed that the brain operated via a sort of natural selection process that he called neural Darwinism or neural selection. In brief, the brain operates as a complex adaptive system. Taken as a whole, it works in much the same way as whole ecosystems that are constantly adjusting to changing conditions via natural selection. Millions of bundles of brain cells (called neurons or neurones) form "maps" of our experiences and these maps are in a constant state of flux and rearrangement, second by second, minute by minute. There are dense webs of non-linear connections between these maps and this ensures that multiple different pathways are available for making new connections. Connections and maps that are used repeatedly grow stronger: something has been "learnt" because those same connections can be quickly made in the future. Connections and maps that fall into disuse fade away. If a part of the brain is damaged, new pathways to achieve impaired functions can often be found. There are exceptions because some parts of the brain come preprogrammed for basic survival and early learning functions—damage to these parts has permanent consequences.

Fundamentally, each person's brain develops its own unique maps and pathways, in response to the specific experiences a person encounters, and the meaning they make of those experiences:

> A baby turtle, on hatching, is ready to go. A human baby is not ready to go; it must create perceptual and all sorts of other categorizations and use them to make sense of the world—to make an individual, personal world of its own, and to find out how to make its way in that world. Experience and experiment are crucially important here—neural Darwinism is essentially *experiential* selection. (Sacks, 2015, p. 361, emphasis in the original)

In a chapter of his autobiography called *A New Vision of the Mind* Oliver Sacks discusses his own reaction to getting his head around Edelman's theory of neural selection (2015). Sacks' studies of patients with neurological challenges are well known—for example, another of his books is called *The Man Who Mistook His Wife for a Hat* (Sacks, 1985). In his autobiography Sacks said that the theory of neural selection made immediate intuitive sense, explaining why obscure and difficult-to-treat neurological conditions might arise. But, by his own admission, he initially found it hard to let go of his classical training in structural brain functioning. In the classical model the brain was "regarded as a collection or mosaic of little organs, each with its specific functions but somehow interconnected. But there was little idea of *how the brain worked as a whole*." (Sacks, 2015, p. 356, emphasis added). This would also be true of how I taught my human biology students about the workings of the brain—I taught the parts but never gave much thought to part–whole dynamics.

I've related this story for a practical purpose, just as Sacks did himself. Our personal experiential maps are hard to remake once they are strongly established. We can anticipate that considerable intellectual effort might be required. There were some indications of this in the discussion of establishing and changing habits (Chapter 9). In Chapter 4 I outlined a range of problematic ways of thinking about the world that need to be remade if we hope to establish a firmly grounded complexity orientation. I found it personally challenging to remake some of my own tacit beliefs/ontological assumptions about how the world "is". Having grown up with tacit assumptions about the linear and gradual nature of change over time, the possibility of abrupt non-linear phase shifts in the natural world was particularly disturbing to me. But once that meaning-making threshold has been crossed, a whole new way of seeing the world emerges and there is no going back.

The idea of "threshold experiences" captures this sense of profound shifts in meaning-making that occur when whole chunks of maps get rearranged. When these thresholds relate to a vocation, they typically lead to a whole new way of *being* that type of worker (Vaughan et al., 2015). Karen Vaughan and her colleagues worked with doctors, engineers, and carpenters, looking at how they learnt "on the job". I think their findings would also hold true for teachers. Classroom work is intense—new experiences come thick and fast and typically require an immediate response. Imagine what is going on in terms of neuronal mapping and adapting, second by second, minute by minute. No wonder learning to be a teacher is so exhausting! No wonder it is really hard to change ways of "being" in the classroom, once they are firmly established.

On a more optimistic note, we could also reflect on the intense gratification that comes when students have "aha" moments. In those moments, something the teacher has said or done helps student(s) rebuild their relevant brain maps in ways that make new sense out of existing ideas and experiences. These moments are so powerful that we tend to remember them as highlights, both of our teaching and in our personal learning. And so I turn next to efforts to leverage known learning challenges to make it more likely that "aha" moments will come.

Constructivism as active meaning-making

Does the term *constructivism* have negative connotations for you? It got a bad rap in the last two decades of the 20th century when it was often mistakenly presented as a teaching approach. It is actually a *learning theory*. Basically it says that each of us must actively construct our own meaning as we learn. This psychological insight strongly aligns with the biological theory of the brain as a complex adaptive system. But it doesn't follow that students should be left to their own devices because only they can do the active meaning-making. As I next illustrate, finding ways to work with the complexity of brain functioning can lead to more effective teaching strategies.

The huge and ever-expanding field of misconceptions research is perhaps the most obvious application of constructivism to teaching. Students' experiences can lead them to build maps that don't quite get relationships correct. This can be obvious and amusing in very young children as they learn to parse and name their world. I'm smiling as I recall a younger cousin who initially called all larger mammals "Georgor" because the family had a cat named George. I also recall a primary-school student who excitedly explained an idea on a reality TV show some years ago. He had made a connection between rubbing up static electricity in his hair and the extrusion lines in magnetic strips around a fridge door. For him the extrusion lines were the hairs that contained the (magnetic) force.

Learning new science concepts is particularly prone to the formation of misconceptions. Our experiences of the world are not necessarily a reliable guide to science concepts, which can be counterintuitive.

As we've seen in earlier chapters, complex systems concepts such as emergence might also seem counterintuitive at first, so it is important that students have opportunities to explore and experience them in meaningful contexts. One of the papers from the initial literature search reported on a meta-analysis of likely misconceptions that can arise when learning about the climate as a complex system (Shepardson et al., 2012). This paper alerts educators to a number of meaning-making challenges that could crop up:

- key climate-change trends have long timescales that are not directly observable, especially when compared with short-term, directly observable weather changes
- complex cause-and-effect interactions can include seemingly opposite effects (for example when an abnormally heavy snowfall is attributed to global warming because of the intensification of storms)
- linear thinking is a hindrance if students are unaware of interconnections among the many components of the system and they do not see feedback loops
- there are multiple ways to set system boundaries (local; regional; global) and these are not necessarily clearly set when a specific example is being discussed
- many students have a simplistic understanding of sources and sinks in the carbon cycle
- the uncertainty inherent in aspects of climate science itself might be misinterpreted as meaning that none of the science has any more authority than other espoused views (as we see in conspiracy theories for example).

How can teachers respond in practical and manageable ways to these complexities of students' meaning-making? It is obviously impossible to carry working knowledge of every possible misconception that students might hold. Well-known learning problems such as those just outlined can provide one useful starting point for noticing and responding to students' meaning-making. Also, look out for resources such as the ARBs which are designed to support assessment for learning. Such resources provide an analysis of the responses made by a range of students when completing a specific learning/assessment activity. The Waterways assessment item in Chapter 12 is one example. The Food Web item in Chapter 4 is another. The research introduced next takes this sort of analysis a step further. Rather than exploring single ideas or tasks, the researchers have explored more generic meaning-making challenges to describe things that teachers need to keep in mind in a range of contexts. All three lines of research discussed next happen to be set in physical-science contexts. Don't let that put you off if you find the thought of reading about physics challenging. I have tried to write about the different research projects in a way that draws out broader implications.

Leveraging the concept of mental models

I have already briefly introduced research by Lemmer et al. (2020) that used complex systems ideas to explore students' understanding of magnetism. Some children have opportunities to play with magnets from a very young age. A strong magnet on a simple dowel and string fishing pole was one of my 4-year old grandson's favourite presents this past summer. He quickly learnt to distinguish "iron metal" from other potential attractants. However, as I briefly outlined in one example above, children can also readily form misconceptions about magnetic force, and about the relationship between magnetism and electricity. Lemmer and his colleagues used this rich experiential context to unravel patterns in young people's meaning-making about magnetism. Their starting point was that our conceptions are *dynamically emergent structures*. Their research described three different types of knowledge structures that interact to form the overall conception of magnetism held by each individual. These were briefly introduced in Chapter 13 and are now outlined in a little more detail in Table 6.

Table 6. Three different types of knowledge structures that our brains construct (after Lemmer et al., 2020)

Type of knowledge structure	Example (context = magnetism)
Intuitive fragments of knowledge: arise from everyday experiences and are weakly structured	The bigger the magnet the stronger the force (this is a misconception)
Structured misconceptions: science concepts become distorted as students attempt to integrate them with their intuitive knowledge	Magnetic field lines are concrete objects All metals are attracted to magnets
Theory-like system of knowledge elements (also known as schema or mental models): these are well established and used reasonably consistently in explanations	Magnetism is electricity (on the way to a correct understanding but not quite there yet) Magnetism is understood as a force field

The idea of mental models or schema is often applied in education contexts. Piaget rightly remains famous for his theories of adaptive changes to children's mental models as they gain greater and more interconnected experiences of the world. What Lemmer et al. have demonstrated here is the dynamic relationship between mental models, individual concepts that are the formal learning focus, and the diverse everyday experiences that all young people bring to their learning. Teachers need to keep all three of these types of knowledge structures in mind when they are designing learning experiences, so that the conceptions they want students to build are more likely to be the ones that do emerge (Lemmer et al., 2020). Earlier chapters have emphasised the importance of metacognitive reflection—this discussion of mental models provides another reason for exploring the complex ways in which learners are constantly making meaning of all their experiences of the world.

Using a systems approach to keep relationships between concepts in clear view

Chapter 4 discussed the challenges that arise when teachers break complex ideas down into simple parts to make them more accessible. If part–whole thinking is neglected, this widespread, well-intentioned practice can actually work against building the capabilities needed for complex systems thinking. There is also some evidence that "keeping things simple" can make target concepts *less accessible* than if a systems approach is used right from the start. My example continues with the context of magnetism and extends to related concepts of energy and force fields.

It is quite common for young children to experience the effects of force fields when they play with magnets, or rub up static electricity to watch what happens. But the underlying concepts of energy transfer and force fields are typically not introduced until the middle-school years, or later. Also, these concepts are usually introduced separately. Each of them is known to be prone to a range of misconceptions—for example many students hold the idea that energy is "used up" when something happens. But some science educators now argue that children can begin to develop systemic insights into how energy behaves as early as Year 2, if a systems approach is used to purposefully bring the concepts of force fields and energy transfer and transformation together (Nordine et al., 2018).

Jeffrey Nordine and his colleagues report that such experiences can be turned into more helpful mental models when the relevant context is treated as a simple double system. In Figure 17, System 1 is the place where some process is causing the available energy store to decrease and system 2 is the place where a different process can make use of the increase in energy. If we see the energy-transforming processes as the equivalent of mechanisms (M), I think there are interesting resonances with the CMP model discussed in earlier chapters and applied to the idea of progression in Table 4 in Chapter 12.

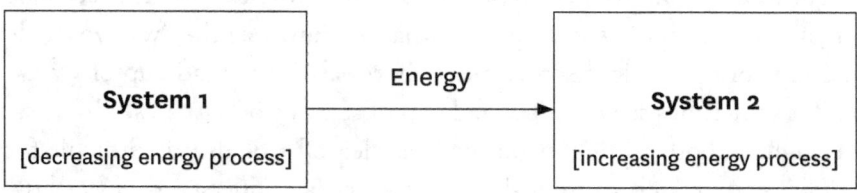

Figure 17. The basic arrangement of an "energy transfer model"
(Nordine et al., 2018, p. 187)

One example given by Nordine et al. is a simple solar cooker. System 1 is the sun, where a thermonuclear process releases energy in the form of light. System 2 is the cooker, where this light energy is concentrated to speed up the movement of food particles, raising the temperature of the system and hence causing the food to be cooked. In between, but invisible to us, is the electromagnetic force field that allows the light to radiate through seemingly empty space.

Notice how similar this diagram looks to a simple stock-and-flow model, as introduced in Chapter 6. In fact I became aware of this body of research when I read a post on the Science Teachers' Facebook page that advocated using stock-and-flow models to teach students about electricity. This prompted me to go hunting for references to research in this area. I chose to use Nordine et al. because of their wider discussion around the explicit use of systems thinking to teach energy concepts in a range of contexts (including but not limited to electricity).

Nordine et al. point out that cognitive tools like the simple model in Figure 17 keep the relationship between different parts of a simple physical system in clear view. This helps make complex ideas more accessible. They also note that this approach can be applied consistently across many different disciplinary contexts. Energy has tended to be treated differently in biology, chemistry, and physics, but this "systems-transfer" approach to teaching the concept of energy works in the same way in all the science disciplines (Nordine et al., 2018).

Embracing complexity to manage the learning of difficult concepts

If basic science concepts such as force fields and magnetism are tricky for younger students to meaningfully incorporate into their mental maps, the challenge is considerably magnified when senior secondary students begin to encounter the concepts of quantum physics. To illustrate, one of the more accessible examples is what is known as the "wave/particle duality" of light. Light sometimes behaves as if it is made up of waves, and it sometimes seems to be made up of streams of particles. Physicists say light is both, simultaneously. This idea of a duality is difficult for students to grasp because it defies our everyday understandings of how objects behave in the world. In fact quantum physics changes the very concept of what an object *is* the world (Levrini & Fantini, 2013).

Levrini and Fantini critique the *hyper-simplicity* of visual images and metaphors typically used in physics textbooks. They acknowledge that such images are intended to make complex ideas more accessible, but their research with Italian senior secondary students suggests that these images actually do the opposite. Simplified visual images are always *partial*. In the case of the wave/particle duality, they can show models of waves, or of particles, but they can't show both at once. They are "dangerous or unproductive simplifications" (p. 1897). Their research suggests that many students end up concluding that quantum physics is a "mess"

(p. 1904) because they can't resolve differences between classical and quantum models of phenomena by drawing on their previous learning or experiences of the world. However, as the title of this subsection suggests, the researchers' solution to this dilemma is to embrace and work with complexity, rather than trying to gloss over or ignore it.

Levrini and Fantini acknowledge the cognitive struggle that scientists faced when they first encountered the need to rethink simple classical physics concepts to arrive at a meaningful understanding of more complex quantum ideas. Students must wrestle with this same struggle:

> Entering modern physical thinking means recognizing how traumatic it has been to give up the classical image of the world, together with becoming aware of the revolutionary contribution of modern physics in redefining very basic thinking categories. (Levrini & Fantini, 2013, p. 1899)

I have already highlighted this problem in the introduction to this chapter, citing Oliver Sacks' struggle to replace ideas of classical neurology with the notion of the brain as a complex adaptive system, with all the implications for his work that followed from that. Levrini and Fantini propose that physics teachers adopt a history of science approach, introducing new ideas in a rich chronological sequence and allowing ample time to explore each version, comparing and contrasting multiple perspectives as these build up.[72] In this way, students see how their own meaning-making challenges were also those of physicists, and how scientists' emergent ideas were able to resolve evidential dilemmas and create richer, more productive webs of understandings. In this "longitudinal" approach:

> The "game" of modelling quantum phenomena is systematically analysed and compared with the models already encountered by students during the study of other physics topics (classical mechanics, special relativity and thermodynamics). (Levrini & Fantini, 2013, p. 1899)

---

72 I debated adding "the history of ideas" as a type of context in Chapter 11 but decided that any examples I could give would add the pro-science bias. I expect that those with deep knowledge in the social-science disciplines could readily think of examples that would illustrate this possibility equally well. I wonder if one potential candidate is the idea of the 'tragedy of the commons' as introduced in Chapter 10.

Notice how the complexity of meaning-making becomes an explicit learning focus here. Misconceptions that students might hold are anticipated, because they have also been those that scientists struggled with in earlier times. There is a strong metacognitive dimension and the learning environment fostered by the teacher is "properly complex territory" (Levrini & Fantini, 2013, p. 1899) through which students are carefully guided.

Building awareness of changes in our own meaning-making

Oliver Sacks has a warning about a potential trap when using the history of science to teach contemporary ideas. He drew on the work of Stephen Jay Gould to warn against dismissing earlier ideas as wrong-headed and generally not very smart. In his book *Wonderful Life* Gould explored the considerable cognitive and practical challenges faced by palaeontologists of different eras as they described the relationship between fossils in the Burgess Shale (some of the most ancient rocks known on Earth) and today's flora and fauna (Gould, 1989).

In the middle of the 20th century, reworking of the research of the early palaeontologists led to the emergence of new ideas about evolution, and specifically the new theory of "punctuated equilibrium". Instead of being seen as primitive precursors of today's species, palaeontologists gradually, and initially reluctantly, came to understand that most of species found in the Burgess Shale did not contribute to ongoing evolution at all—they had become extinct as part of some calamitous ecological change at the dawn of the Cambrian geological era. The new theory emerged with difficulty because the researchers who were re-describing the fossils had to think *against* assumed practical and theoretical wisdom of the time. They literally had to remake their own cognitive maps. They were supported in this struggle by other palaeontologists who were wrestling with the idea that a later, similarly calamitous, ecological change caused the extinction of the dinosaurs. Evidence for a different, complexity-oriented, way of thinking about the fossil record was growing and becoming more convincing (Gould, 1989).

I recall first reading *Wonderful Life* when I was still a secondary-school biology teacher. I took from it an enlarged understanding of evolution but I totally missed the strong "epistemic" thread—the discussion of the politics of knowledge and how what we already "know" influences what we can see and think. The influences that are most widely held in our society are those that are most likely to be invisible to us. Encouraged by Oliver Sacks I reread *Wonderful Life* in the summer of 2019/2020 and took from it a whole new set of ideas about the importance of respecting the complexity of cognitive struggles faced by those who endeavour to build new knowledge for the benefit of all. I also recognised early seeds of my own struggle to engage with the non-linearity and capriciousness of punctuated changes in life on earth, although conscious acknowledgement of that understanding was still several years in the future when I first read the book.

In Chapter 10 I introduced *Why Reading Literature in School Still Matters*. In this book Dennis Sumara argues that it is important to support all students to read literary texts (Sumara, 2002). He specifically writes about the complexities of cognition, and of teaching. He encourages teachers themselves to reread personally significant texts at intervals of some years. Each reading will be different because the reader is quite literally a different person, given their meaning-making experiences in the intervening years. I hope I have convincingly demonstrated this in my own thinking the paragraph above. It seems to me that this also provides a strong argument for a spiral curriculum design, where significant ideas are purposefully revisited at periodic intervals and remapped to other new ideas and insights that have been acquired in the meantime.

Finally in this chapter I turn to a more recent research project. Like Sumara, Shae Brown explores teaching strategies to build students' awareness of complexity. He too is conscious of the role that students' own thinking plays in constructing new meanings from their lived experiences. The term *complicity* is used to signal this sense of the complex two-way nature of meaning-making. Just as the world imposes its meanings on us, so our existing ideas and our actions shape the meaning of our experiences in the world we encounter. With secondary-school students, Brown uses the ubiquitous and seemingly simple concept of time as his starting point for exploring this complex idea.

Exploring time as a complex concept

> It might be considered unusual to be including cognition in the teaching and learning of complexity competence, yet the issues we face in the 21st century require us to not only understand complex phenomena generally, but to understand our relationship with and within particular phenomena. (Brown, 2019, p. 7)

Only some students will have opportunities to learn quantum physics. *All* students experience "time" as part and parcel of daily life. Brown (2019) explains how he uses this simple and ubiquitous concept to explore the complexities of cognition and hence develop students' complexity competence.

Brown argues that secondary-school students are very aware of the complexities and uncertainties they experience in their daily lives. He says that they experience a dissonance between this reality and what happens to them at school, when structures and practices typically assume a linear, deterministic simplicity. As one example, assessment for school-exit qualifications often presents students with an unforgiving linear experience of their destiny as learners. In traditional examination systems they come to a specific finish-line or cliff in their school learning life when they are judged to be a success or not, regardless of all that has gone before and all the new learning experiences that might follow. In this way, familiar conceptions of time can render both teachers and students complicit in building some students' identity as unsuccessful learners (Brown, 2019).

When students' *experience of time* is explored in a more complex framing, these familiar practices can lose their power to define students' futures. Students gain "temporal breathing space" (Brown, 2019, p. 19) to construct a different set of relationships between their past and possible futures. This sense that students can remake their future identities as learners can also be found in learning-to-learn models created by other research teams who assume a complexity perspective for their work. For example, Figure 18 shows a schematic model of learning-to-learn created by Ruth Deakin Crick and her colleagues during a long-running research programme (Deakin Crick, 2014). Notice how they symbolically position identity and a personal sense of purpose at the very heart of learning-to-learn. Notice also that this model uses the metaphor of "learning power", which was introduced in Chapter 9.

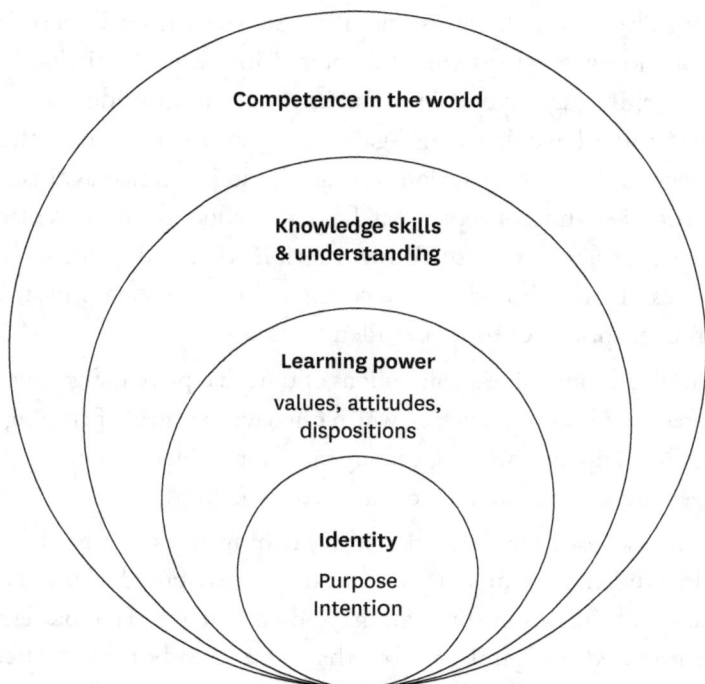

Figure 18. A summary model of learning-to-learn as a complex process
(After Deakin Crick, 2014, p.16)

Brown uses what he calls a "patterning" approach where students work with visual models based on biological metaphors. Concentric spheres, as in Figure 18, are one type of pattern he uses. However he begins with an exploration of time experienced as a spiralling pattern of events. There are illustrative examples of spiral rhythms at all levels of scale in nature itself: day and night; seasons; lunar cycles, generational or life cycles, and so on. The spiral nature of students' days and weeks provides another set of experiential examples to layer on to spiral patterns in diagrams of increasing complexity. Students begin to talk openly about differences between earth time, clock time, and school time. They can also acknowledge differences in experienced time in different contexts.[73]

---

73 My friend and ex-colleague Karen Vaughan recently introduced me to the idea of "trauma time". This is experienced as a sensation of time standing still but what happens during that experience of hiatus can be revisited as vivid flashbacks for years to come. Sheldrake (2020) says that our sense of smell can act as a trigger for flashbacks to traumatic experiences.

Brown also notes that this spiralling sense of time is prevalent in indigenous knowledge systems. Chapter 8 introduced this idea in the context of thinking beyond human timescales to consider our obligations in terms of whakapapa. Again, making a direct comparison of worldviews comes with a caution. Just as Gould cautioned against seeing science theories and concepts that have subsequently been replaced as not very smart (see above) so Brown cautions that Indigenous students' experiences of time should not be consigned to a mistaken past, before Western conceptions of time prevailed:

> We need to counter linear conceptions of time that place Indigenous cultures as an historical artefact, which can have the effect of creating 'temporal displacement', disappearing the complex identities of Indigenous students and their cultures. (Brown, 2019, p. 21)

Brown also uses the metaphor of branching trees or mycelia.[74] This can help students to model non-linearity, distributed causality, self-organisation, bifurcation (branching) and emergence. Tree patterns can be superimposed on spiral patterns that have already been created. He also uses the idea of "seed patterning" to introduce students to challenging conceptual ideas about phenomena that teeter on the edge of chaos, while also being self-organising. A whole new and more complex future can have its origins with a tiny seed. Inside a seed (or a pupa for that matter) the available material is transformed into something new and different, sometimes suddenly and sometimes slowly. Brown says that this metaphor allows turbulence and change to be experienced as creative and generative rather than inherently threatening. In this way, students' explorations of time support them to build dispositions to tolerate uncertainty. As we have seen repeatedly in the preceding chapters, is an important dispositional component of complex systems thinking.

---

74 The branch-like threads that form the body of a fungus are called its mycelium. Fungi are further discussed in Chapter 15.

Chapter 14 reading guide

This chapter explores *learning* through a complexity lens. Understanding cognition as a complex phenomenon can be helpful for finding strategies to support students when they face a range of learning challenges—or when they need to be challenged and extended. Relevant insights range from the micro-level (neurobiology) to the macro-level (ecologies of learning).

1. Do you agree that all students should have opportunities to learn about the complexity of learning? If yes, how might this topic best be fitted into the curriculum, so that it is available to everyone? How could teachers be supported to do it well?

2. Can you think of a powerful "threshold experience" that prompted you to see yourself as a teacher in a whole new light? Why was this moment so significant for you? Can you trace slower but profound changes in your thinking over time, like the ones I have included from my own personal experience? What prompted you to reflect on this journey in ways that led to personal insights about the changes?[75]

3. Many teachers avoid including "wrong" ideas in the learning experiences they offer to students because they do not want to confuse them. What do you think of the argument that older students, especially, should learn complex ideas via an exploration of the cognitive struggles of past scholars and researchers? Can you think of relevant examples in your own discipline area(s)?

4. In what ways does the term *complicity* reflect the both/and thinking that is central to complexity theories? Do you agree with the argument that all schools and all teachers are complicit in students' identity formation as learners? Why or why not?

---

75 In my case, for example, rereading significant books at an interval of some years was an important influence.

Chapter 15

Complexity at work in the world

Complex systems are everywhere, at multiple levels of scale. The more researchers explore, the more complexities they find. New insights into complex systems raise important existential questions about how we should live on the only planet we call home. Every student is entitled to gain a sense of the wonderful life that surrounds us, and to bring these new sensibilities to their participation as informed and engaged citizens. Many students will also need a basic level of systems literacy to take into higher education or their chosen work pathways.

I originally thought about this last chapter as the "pathways" chapter. Any teacher might ignite a spark of inspiration about a possible future career for a student (Vaughan, 2010). Therefore, I envisaged a chapter that outlined a range of career opportunities which require students to be or become accomplished complex systems thinkers. I thought teachers might find a range of examples useful, should opportunities for career-focused conversations crop up in the classroom. There are elements of this pathways focus threaded through the chapter but, looking back across the whole learning journey set out in the book, I now see that focus as too limited to be the last word.

For starters, there is a clear and consistent message in the research literature that every citizen needs to be or become a capable complexity thinker if they are going to fully participate in confronting the challenges that face us in this century. It would be wrong to conclude that this type of thinking will be more important for some students than for others, on the basis of potential employment needs.

Secondly, I wanted to ensure that the examples I chose would not all adopt a view from outside systems looking in—a challenge that still lay in the future as I embarked on the book. I still find it hard to think beyond this traditional framing of knowledge unless I pause and do so deliberately. The habits of a lifetime don't change overnight! To try and think from that "inside systems" perspective I pondered what it *might mean* for young people to take an acute sense of the complexity inherent in the world into their wider lives beyond school, including but not limited to potential employment.

That question about meaning directed my wondering towards the *aesthetic* dimensions of learning. Might complexity thinking enrich our experiences of the world in unique and wonderful ways?[76] It's interesting that aesthetic considerations are more likely to be raised in subjects that tend to be marginalised in the traditional school curriculum. I am thinking of the arts in particular. What might learning look like if the amazing complexity and wonder of our world was a focus right across the curriculum? I can only scratch the surface of an answer in what follows.

Complexity thinking is important for teachers

If learning to think in terms of complex systems is seen as important for every student, it follows that all teachers need to have these capabilities themselves. You can't teach what you don't know. I don't mean to sound glib—the argument is obvious, but this is a challenge that should not be underestimated. In earlier chapters I have indicated that my own learning journey has been a bumpy one that has taken place over a number of years and is still ongoing. A clockwork frame of reference for how the world works is so ubiquitous that it is not easily displaced. Besides, there is a chicken-and-egg dilemma. Who will teach the teachers so that future generations of teachers are already able to think in complexity terms before they confront the many other challenges of learning to be a teacher?

---

76 This sense is present for example in the title Stephen Jay Gould chose for his book about the fossils of the Burgess Shale, *Wonderful Life* (see Chapter 14).

What might motivate current teachers to want to embark this learning journey, when there are so many other urgent things to attend to in the course of their working lives? One research team makes the case that there are work-related benefits in coming to understand learning itself as complex, emergent, and long term, as opposed to traditional linear, short term, and "reasonably straightforward" (Jess et al., 2018, p. 466). This teacher educator team found that their own growing understanding of complexity helped with building a deeper understanding of learning needs of their student teachers. They said they became "more adaptive professionals" (p. 466). The examples in Chapter 14 support this point. Several of the research papers cited there argue that deliberately taking a complexity approach to tricky topics can actually make learning easier for students. Another potential benefit is that the visual approaches used to teach complexity are inclusive—students with specific learning needs can access more of the curriculum than they might otherwise be able to do (see chapters 7 and 9).

I can also see potential benefits associated with the current move towards making New Zealand's curriculum more genuinely bicultural. As just one indication of this shift, the draft histories curriculum requires that difficult events in our past are explored from multiple perspectives (as per the example in Chapter 11). The capability for taking different perspectives is also centrally important in complexity thinking (see Chapter 7) so there is a potentially useful overlap in learning goals right there. But I think the ultimate value lies in a deepening understanding of complex systems thinking as a way of knowing. As Chapter 8 outlined, some features of complexity thinking differ from more traditional Western ways of knowing and are more akin to indigenous knowledge systems. Here a second indication of the shift to a more bicultural curriculum comes into view. I am thinking about the requirement to include elements of mātauranga Māori in the exemplars developed to support the new NCEA achievement standards used for assessment in the senior secondary school. In my view, the two knowledge systems cannot be safely juxtaposed in any subject unless and until they are actually understood to be quite different knowledge systems. Complexity thinking can help with that, by drawing attention to features of Western knowledge that might otherwise remain invisible to many teachers.

What about aesthetic considerations? Given the uncertainty inherent in complex systems, I wonder if there is a less tangible benefit in the need to slow down and spend more time exploring "it depends" contingencies (see Chapter 5). Even very simple systems such as that introduced in Figure 2 (Chapter 2) could trigger rich, inclusive conversations that draw on learners' experiences of the uncertainties inherent in life outside of school. Many learners experience a tension between these lived uncertainties and the more certain knowledge seemingly valued in school (Brown, 2019) so some of the value could lie in simply acknowledging this reality. As I will shortly outline, complexity thinking could also support teachers to ponder previously unthinkable questions about us and our place in the universe.

Complexity thinking underpins many management strategies

Moving from teacher learning to student learning, my original "pathways" plan for this chapter now comes to the fore. I draw on a specific example where the school-to-work transition is clearly in focus, to explore the value that complexity thinking might add to students' overall learning experience. Very recently, I had the privilege of visiting the Business Academy of Manurewa High School in South Auckland. It is possible to read a complexity framing into the mission statement of the academy:

> By connecting students, alumni, business and tertiary educators through partnerships, the MHSBA aims to transform the lives of MHS students, and through them, the lives of their families and communities. (https://www.manurewa.school.nz/curriculum/business-academy)

How exactly might learning experiences orchestrated by the academy teachers and their business partners *transform* lives? There is a material argument to be made—gainfully employed (and employable) young people will be less likely to struggle financially. This is important, but the challenge I have set myself is to also explore the wider potential of these experiences if the learning is framed in terms of complex systems dynamics.

As I met with several of the teachers and listened to the experiences of a Year 13 student, I could detect a certain deliberate leveraging of complexity principles, at least on the part of one of their main business partners, Mainfreight. The student was taken aback when she first encountered the saying "if you are on time, you are late." Mainfreight organises the workforce in their transport/logistics business as teams that work tightly together and can swop in and out of one another's roles. In this environment, prospective employees need to learn early that what they do impacts the whole team. They are inside a system that has been carefully engineered for both efficiency and resilience.

This discussion reminded me of a conversation with Karen Vaughan in which she made a distinction between being "on time" and having a sense of "timeliness". The former could be about rule-following (e.g., not being late). The latter is about having a sense of the impact of your own decisions and actions within a wider system, and being able to support the goals of the collective by making timely decisions. One example might be anticipating the impact of a mistake or of something breaking, and acting swiftly to right the situation instead of waiting for someone else to act, and possibly for chaotic change to unfold. Along these very lines, the student also told us about an episode where a numerical mistake had been deliberately inserted into one part of a team exercise. First, the problem had to be noticed, and then a strategy to solve it had to be worked out. This too had proved to be a memorable piece of learning inside a complex system.

Playing the devil's advocate, I could argue that the learning would still be valuable regardless of a student's actual awareness of the complexity principles on which the work system is organised. One immediate and prosaic counterargument might be that it would be difficult to move beyond the team worker level—to a supervisor role, say—without an understanding of the principles behind the organisational structure. More immediately, my sense is that even a basic level of complex systems thinking would help to avoid a feeling of being at the mercy of capricious forces you cannot control (e.g., never understanding why it's not OK to be just on time). To echo David Perkins' point in his book about the importance of "whole" learning experiences, workers could feel more in control once they understand "the hidden game" of the power structures within which they are immersed (Perkins, 2010). With that thought, I turn next to one of the most powerful yet largely taken-for-granted structures that has the potential to play hidden games with all our lives.

Complexity dynamics can be used to check claims to truthfulness

> ... it is imperative for students to understand how the Internet shapes the information they receive. (McGrew et al., 2018, p. 166)

The internet is a vast adaptive system. It is constantly learning as we interact with it. Its services are so ubiquitous and useful that we tend to take its presence for granted. While our personal uses are likely to be benign (from our perspective), many hidden games and power structures are constantly at play. I personally know nothing about the so-called dark web except that it exists and is used for criminal activities. What about the less dramatic, but no less damaging, power games played by disinformation experts? These shadowy players leverage the internet's adaptive features to promote agendas that are not in our best interests, even though those who are hooked into them might sincerely think they are. Learning about these hidden games could be part and parcel of many different learning contexts, as I next outline.

Several years ago, American history educators Sam Wineburg and Sarah McGrew conducted a small experiment. They chose two similar looking websites, one that set out reliable information and one that contained deliberate misinformation. They asked different groups to work out which was which. The undergraduate and PhD students tried to use critical thinking and were not very successful at spotting the fake content. The other group were employed as professional fact checkers. After a quick scan of the content, they clicked away from the pages they had been given, searching out the ecology of their connections across the wider internet. (The researchers call this "lateral reading".) They went back and read the content once they were reasonably certain they knew which information was genuine. Needless to say, the professional fact checkers were more successful in spotting the fake material than the students had been, and they took considerably less time to arrive at their conclusions. The researchers pointed out that some communicators with a hidden agenda will use a range of ruses to make their content look trustworthy. For this reason, being taught to evaluate content using critical-thinking skills is not a reliable strategy on its own (Wineburg & McGrew, 2017).

Lateral reading is a technique that can be taught, at least to older students.[77] They need to learn to ask three main questions. Who is behind the information? What is the evidence? What do other sources say? (McGrew et al., 2018, p. 168). A certain level of understanding of the internet as a dynamic complex structure should give students an appreciation of why each of these questions is important, and why all of them do need to be asked and answered. Such understanding won't necessarily make lateral reading any easier to actually carry out, but I wonder if the dispositions fostered by complex systems thinking would be helpful in this context. Students who have built systems-thinking dispositions (Chapter 9) might be more habitually inclined to slow down and check information in this way.

The research I have just introduced was motivated by the concern that students need to be better prepared for civic participation online (McGrew et al., 2018). Some manipulation of potential civic participation is widely known. I'm thinking here of the Cambridge Analytica scandal, for example, where Facebook data was mined and used in ways that were intended to influence the outcomes of elections in several nations.[78] My choice of Wikipedia as a reference is deliberate and ironic. I recently listened as complexity scientist Markus Luczak-Roesch talked about his research with a group of secondary-school economics and business teachers. He introduced the challenge of detecting manipulation of Wikipedia entries to reflect a particular political perspective. Mathematical modelling techniques developed by his team can detect subtle, non-random patterns in editing activities. The modelling can pinpoint systematic manipulation of related entries, carried out in a covert manner with the intention of avoiding detection. Once a set of entries has been highlighted in this way, human editors can check and correct as necessary.

---

77 Is there another chicken-and-egg dilemma here? How will we know at what stage of their development students can learn to do this successfully without giving a wide range of individuals the chance to try? The experiences of the citizen-science teachers (Chapter 11) suggest that some primary school students might well be capable of working with questions that critique information if they are carefully supported to do so in contexts that are accessible to their current life experiences.

78 https://en.wikipedia.org/wiki/Facebook%E2%80%93Cambridge_Analytica_data_scandal

Most recently, Markus has been tasked to lead a team to work on The Veracity Mission of the National Science Challenge Spearhead "Science for Technological Innovation" (Science for Technological Innovation, 2021). This will use principles and technologies from complex adaptive and self-organising computer systems to check the truthfulness of claims that entice us to choose certain products over others. For example a specific food might be said to have been produced sustainably. There is likely to be a long value-chain, with multiple steps between farm and consumer, so it is difficult for us to check if this is really so. As another example, a product that makes use of indigenous knowledge might claim to be ethically supported by the Māori owners of that knowledge. It would be hard for individual consumers to check that appropriate indigenous governance and oversight really are in place. Up until now, we consumers have had to rely on centralised systems and processes that check such claims. But this is problematic because a centralised approach favours the bigger market players. The team on the Veracity Mission aims to leverage their knowledge of complex adaptive systems to decentralise processes in ways that "level the playing field for all involved" (Science for Technological Innovation, 2021, n.p.). As one specific example, Māori researchers on the team are looking to develop a system of digital tags for traditional knowledge and biocultural products. The research aims to develop technological systems that will allow all relevant participants to check claims of products that aspire to carry the tags.

This subsection on complexity dynamics and truthfulness took shape around the general theme of the complex relationship between information technologies and truthfulness. There are several types of career opportunities here. For example some students might aspire to become professional fact checkers. A citizenship level of systems literacy would arguably be necessary for this work (see Plate & Monroe, 2014). Those who aspire to become mathematical modellers or complexity scientists will need to develop the "advanced" level of Plate and Munroe's typology of levels of systems literacy (Chapter 13). Some of this learning can doubtless wait until university, but secondary-school science and mathematics teachers must lay the critical foundations, and hopefully spark the interest of science-able students in research careers that require complexity capabilities as a bottom line.[79] The many examples throughout the book have suggested that these opportunities are varied and challenging. My final examples illustrate how accomplished complexity thinkers can push the limits of our imagining further than we might have ever thought possible.

Complexity thinking invites the "unthinkable"

The subtitle here references a distinction made by sociologist Basil Bernstein between practical, utilitarian (thinkable) knowledge, and knowledge which is "more abstract, esoteric and open-ended" (Hughson & Wood, 2020, p. 6). The contrast with practical knowledge is made by labelling this other type as unthinkable knowledge. Hughson and Wood draw on Bernstein's insights to explain that unthinkable knowledge "has the capacity to take individuals beyond their current realities and allows them to imagine new and different realities" (Hughson & Wood, 2020, p. 6). They make this argument in support of the important role of disciplinary knowledge within the curriculum. I agree with this argument but would also extend it a step further. Complexity thinking transcends the separate disciplines, drawing the unthinkable knowledge generated by each of them into a bigger, more dynamic whole (see also Chapter 10). I use two examples to illustrate this argument. Many more could be proposed if space allowed.

As I was working on this chapter two Australian geography researchers wrote a blog post that challenged readers to imagine the Darling-Murray river basin as a complex system with no human presence. What would we see?

> We would see a natural process of river expansion and contraction, of rivers doing exactly what they're supposed to do from time to time. We'd see them exceeding what we humans have deemed to be their boundaries and depositing sediment across their floodplains. We'd see reproductive opportunities for fish, frogs, birds and trees. The floods

---

79 Students interested in such careers could be directed to explore the website of Te Pūnaha Matatini: https://www.tepunahamatatini.ac.nz/. This Centre of Research Excellence (CoRE) brings together complexity researchers from a range of New Zealand institutions to work together on complex investigations. Perhaps one of the most widely known at the moment is the work of the mathematical modelling team, used to support government decision-making during the COVID-19 pandemic.

would also enrich the soils. Floods can be catastrophic for humans, but they are a natural part of an ecosystem from which we benefit. (Parsons & Thoms, 2021)

Several weeks prior to this posting a devastating flood swept through the Darling-Murray riverine system—at least it was devastating from a human perspective. But as this quote illustrates, flooding is a natural part of such systems and a source of renewal for all the other life forms on the riverine plains. Here human interests come into direct conflict with those of other species. The researchers point out that linear management strategies, such as building higher levees, are bound to fail. They also severely disrupt the natural functioning and resilience of riverine ecosystems. They say we need new approaches that "combine community participation with research, resilience and adaptation to make long-term decisions about the future of these complex social-ecological systems" (Parsons & Thoms, 2021). Thinking the unthinkable (a system that carries on without us) has allowed these geographers to reframe "the problem" and propose new ways forward, at least in principle. Importantly, their proposed solution requires the active participation of everyone living on, or impacted by, the health of the flood plains. A citizen-level of systems literacy is implicated for all participants.

Knowledge from a range of disciplines comes together in the work of these geographers: human geography (land uses etc.); physical geography (relationship of rivers and flood plains); ecology (dynamics of life on the flood plains) and of course, complexity knowledge itself. Armed with the latter, Parsons and Thoms were able to imagine the unthinkable—inviting people deeply immersed inside a complex system to imagine it without any human presence at all. Doing this allows the complexity of the overall challenge to come into clearer view, hopefully affording a new approach to solving the challenges raised. No doubt this approach could initially be experienced as very confronting by those with a vested interest in continuing to live on, and draw their livelihood from, riverine plains. My final example is potentially even more confronting because it challenges all of us to rethink our assumptions about ourselves as individuals and as a species.

As this book was nearing completion, NZCER's publisher David Ellis suggested I read a recently published book called *Entangled Life: How*

Fungi Make Our Worlds, Change Our Minds, and Shape Our Futures (Sheldrake, 2020). This book is full of fascinating insights into the largely unnoticed activities of fungi, in particular their complex interactions with plants, with other microbes (algae, bacteria etc.), and with us. Notice the non-linear sense of time in the short quote that follows. It is an important indicator of thinking that is more attuned to indigenous knowledge systems than to traditional Western science:

> plants and mycorrhizal fungi enact a collective flourishing that underpins our past, present and future. We are unthinkable without them, yet seldom do we think about them. And our neglect has never been more apparent. It is an attitude we can't afford to sustain. (Sheldrake, 2020, p. 138)

The insights presented in *Entangled Life* have been generated via two main fields of biological inquiry. Unlike the research into truthfulness outlined above, these inquiry areas do not use complexity methods per se. More traditional scientific methods of inquiry are outlined. Complexity plays its part by opening up previously unthinkable conclusions about the activities of these "lower" life forms. And these conclusions, in turn, open up previously unthinkable questions about ourselves and our place in the universe—at least if your starting point is traditional Western knowledge (see Chapter 8).

The first of these research areas is especially confronting to our assumption that "intelligence" resides only in vertebrates with clearly visible structures we call brains. As Merlin Sheldrake eloquently documents, multiple research projects have clearly established that the thread-like mycelium, which makes up the body of a fungus, demonstrates sophisticated problem-solving behaviours. It might seem unthinkable that "brainless organisms outside the animal kingdom" (p. 16) can be intelligent, yet this is clearly the case in fungi and in other species such as cephalopods (octopi; squid). The metaphor of the great chain of life represents humans as the pinnacle of intelligent species on earth.[80] Although it is largely invisible in our day-to-day thinking, assumptions about our species' superiority underpin almost everything we do and think, at least from the perspective of Western thought systems. It is quite unthinkable

---

80 *Wonderful Life* (Gould, 1989) has an illustrated discussion of the metaphors we use to indicate our relationships with other living things

that intelligent behaviour can be demonstrated by organisms we typically describe as "lower" forms of life. Barriers to our sense of self in relation to other living things are created by these preconceptions, which is perhaps why knowledge from the second of the new research fields described is equally confronting.

New technologies have allowed researchers to study microbes in far more detail than was previously possible. The more they discover in this comparatively new field of research, the more apparent it becomes that all life on earth, including ourselves, is thoroughly entangled with microbes. Quite literally, we could not exist without them. Each of us has our own microbiome. As Sheldrake puts it, "for your community of microbes … your body is a planet" (Sheldrake, 2020, p. 18). Sheldrake is not alone asking challenging existential questions about where "we" end and our environment begins. The complexity of human–environment interface, and of what actually constitutes a "human" cell, has been a matter of debate in scientific communities for at least a decade now. One of my personal favourites is an essay called "The Teeming Metropolis of You" (Buhler, 2012).

More sophisticated research techniques have opened up insights into the complexity of mutualistic relationships between different species. A whole chapter of *Entangled Life* is devoted to lichens. Although a lichen appears to be one plant-like organism, it is actually a complex structure with interdependent fungal and algal components. Both the fungus and the alga benefit from living as one structure so this is called a symbiotic relationship. Notice how existential questions come up again in response to new insights about this complexity:

> Lichens are places where an organism unravels into an ecosystem and where an ecosystem congeals into an organism. They flicker between 'wholes' and 'collections of parts'. Shuttling between the two perspectives is a confusing experience. The word 'individual' comes from the Latin, meaning un-dividable. Is the whole lichen the individual? Or are its constituent members, its parts, the individuals? Is this even the right question to ask? Lichens are a product less of their parts than of the exchanges between those parts. Lichens are established networks of relationships; they never stop lichenising; they are verbs as well as nouns. (Sheldrake, 2020, p. 99)

We have encountered part/whole and both/and thinking already. For example light behaves as both waves and particles simultaneously (Chapter 14). What new questions come into view if we think about ourselves as simultaneously "us" (human cells) and our microbiome, constantly exchanging things between our undividable parts? This isn't actually a rhetorical question. Many researchers have been exploring the implications of this previously unthinkable field of knowledge. For example one line of research at the Liggins Institute, tagged as the Gut Bugs project, is exploring the impact of our microbiome on the health of our digestive systems. Their outreach programme has developed resources to support teachers to introduce this topic in their classrooms:

> It is becoming increasingly evident that interactions between humans and microorganisms have a significant and under-explored impact on many aspects of our health and well-being. The Gut Bugs Trial is a world first clinical research study investigating links between the gut microbiome and health issues such as obesity and diabetes. LENScience and the Gut Bugs Trial team have collaborated with teachers to develop resources that offer opportunities for learning that is contextualised in exploration of this ground-breaking scientific research. (https://www.lenscience.auckland.ac.nz/en/about/teaching-and-learning-resources/gut-bugs-exploring-the-human-microbiome.html)

Once more with feeling: In conclusion

Once we see ourselves as indivisible from our personal microbiome, it becomes untenable to maintain a traditional stance of being outside systems looking in. At all levels of scale, "we" are indivisible from the systems that sustain and support us. All young people need to develop this awareness and many of them will not get there without explicit help and support from their teachers.

Sheldrake poses yet another existential challenge when he discusses our interactions with fungi. Some of them are expert at manipulating the behaviour of other species via aromatic compounds (e.g., truffles), some manipulate the nervous and muscular systems of other species via psychotic chemicals (e.g., magic mushrooms). His question is this: when we take advantage of fungi for our own ends (as we see it) are we using them or are they using us? This question is unthinkable while we continue to see ourselves as outside systems looking in. But it is an important question to ponder because it brings into view the unthinkable challenge that the planet could carry on without us.

I am going to give the almost final word to David Attenborough. He has spent a lifetime documenting the wondrous complexity and interconnectedness of life on our planet. His documentaries are rich in aesthetic appreciation and close place-based observation. He gently poses the same unthinkable question at the conclusion of his most recent book. Talking about the naming of current geological time as the Anthropocene he says:

> What for geologists was a name produced by scientific routine has now, however, become to many others a vivid expression of the alarming change that now faces us. We have become a global force with such power that we are affecting the entire planet. The Anthropocene, in fact, could prove to be a uniquely brief period in geological history and one that ends in the ultimate disappearance of human civilisation. (Attenborough, with Hughes, 2020, p. 216)

On a more hopeful note, his final word is this:

> We can yet make amends, manage our impact, change the direction of our development and once again become a species in harmony with nature. All we require is the will. The next few decades represent a final opportunity to build a stable home for ourselves and restore the rich, healthy and wonderful world that we inherited from our distant ancestors. Our future on the planet, the only place as far as we know where life of any kind exists, is at stake. (Attenborough, with Hughes, 2020, pp. 220-221)

I hope it is clear that more than will is required, important as that may be. It is urgent that we begin to purposefully foster awareness of complexity dynamics, along with systems thinking capabilities and dispositions, as an entitlement for all school-age learners. As the earlier chapters of this book attest, the tools and know-how are freely available to every teacher who is willing to start out on this interesting roller coaster of a personal and professional learning journey.

Chapter 15 reading guide

There are complex systems everywhere and at multiple levels of scale. The more researchers explore, the more complexities they find. New insights into complex systems raise important existential questions about how we should live on the only planet we have to call home. Every student is entitled to gain a sense of the wonderful life that surrounds us, and to bring these new sensibilities to their participation as informed and engaged citizens. Many students will also need a basic level of systems literacy to take into higher education or their chosen work pathways.

Congratulations! If you are reading this guide I assume you have worked your way through the book. While you have been doing so, I hope you have also been on the lookout for other accounts of complexity at work in the world—that your radar is up for new examples and implications. For example, as I was working on these guides, several months after completing the main manuscript, I came across an article about the complexity of New Zealand's braided rivers.[81] This piece really resonated with the research in the Darling-Murray riverine system introduced in this chapter, with the added element that braided rivers have a unique layer of complexity, on top of the general dynamic complexity of all river systems. These were new insights for me—I was not aware that braided rivers are not common elsewhere, or that they are so complex.

1. What new examples of complexity at work in the world have you become aware of since you began this reading journey? Why did they resonate with you?

2. Do you agree that being able to think in complex systems terms should become part of every teacher's intellectual and pedagogical toolkit? Why or why not? If yes, how can we begin to make the necessary changes happen?

---

81 Mitchell, C. (2021). Revenge of the 'zombie' rivers. The Forever Project, Poipoia te Ao, Issue 6, June 2021, pp. 4–8. Also available in podcast: https://interactives.stuff.co.nz/the-long-read-podcast/2021/06/zombie-rivers-canterbury-floods-braided-rivers-farming-irrigation-stopbanks/

3. Thinking back across the whole journey of this book, what has been your most powerful "aha" moment? (I hope there were more than one or two!) How might you personally respond to these new insights, in the short term and in the longer term?

References

Aitken, V. (2021). *Real in all the ways that matter: Weaving learning across the curriculum with Mantle of the Expert.* NZCER Press.

Anderson, D., Buntting, C., & Coton, M. (2020). Using online citizen science to develop students' science capabilities. *Curriculum Matters, 16,* 38–59. https://doi.org/10.18296/cm.0047

Argyris, C. (1970). *Intervention theory and method: A behavioral science view.* Addison-Wesley.

Aronson, D., & Angelakis, D. (n.d.). *Step-by-step stocks and flows: Improving the rigour of your thinking.* The Systems Thinker. https://thesystemsthinker.com/step-by-step-stocks-and-flows-improving-the-rigor-of-your-thinking/

Attenborough, D., with Hughes, J. (2020). *A life on our planet: My witness statement and a vision for the future.* Witness Books.

Benson, T. (2020). *Developing a systems thinking capacity in learners of all ages.* Waters Foundation. https://ttsfilestore.blob.core.windows.net/ttsfiles/developing-st-capacity.pdf

Bernier, A. (2018). How matching systems thinking with critical pedagogy may help resist the industrialization of sustainability education. *Journal of Sustainability Education, 18.* http://www.susted.com/wordpress/content/how-matching-systems-thinking-with-critical-pedagogy-may-help-resist-the-industrialization-of-sustainability-education_2018_09/

Boell, M., & Senge, P. M. (2016). *School climate and social fields—An initial exploration.* https://www.garrisoninstitute.org/wp-content/uploads/2016/05/School-Climate-and-Social-Fields.pdf

Boell, M., & Senge, P. M. (2019). *Introduction to the Compassionate Systems Framework in schools.* Abdul Latif Jameel World Education Lab, Massachusetts Institute of Technology. https://jwel.mit.edu/sites/mit-jwel/files/assets/files/intro-compassionatesystemsframework-march-2019_0.pdf

Bolstad, R. (2020). *Opportunities for education in a changing climate: Themes from key informant interviews.* New Zealand Council for Educational Research. https://doi.org/10.18296/re006

Bolstad, R., & McDowall, S. (2019). *Games, gamification, and game design for learning: Innovative practice and possibilities in New Zealand schools.* New Zealand Council for Educational Research.

Boyle, J. (2018, December 12). The fight to keep ideas open to all. *The Economist.* https://www.economist.com/open-future/2018/12/12/the-fight-to-keep-ideas-open-to-all

References

Bradley, D. T., Mansouri, M. A., Kee, F., & Garcia, L. M. T. (2020). A systems approach to preventing and responding to COVID-19. *EClinicalMedicine*, *21*, 100325. https://doi.org/10.1016/j.eclinm.2020.100325

Brown, S. L. (2019). A patterning approach to complexity thinking and understanding for students: A case study. *Northeast Journal of Complex Systems*, *1*(1). https://doi.org/10.22191/nejcs/vol1/iss1/6

Buhler, D. (2012). The teeming metropolis of you. In T. Folger (Ed.), *The best of American science and nature writing* (pp. 3–7). Houghton, Mifflin, Harcourt Publishing.

Bull, A. (2011). *Library of experiences* [Working paper]. New Zealand Council for Educational Research. https://www.nzcer.org.nz/system/files/Library%20of%20experiences2_1.pdf

Clark, S., Petersen, J. E., Frantz, C. M., Roose, D., Ginn, J., & Rosenberg Daneri, D. (2017). Teaching systems thinking to 4th and 5th graders using Environmental Dashboard display technology. *PLOS ONE*, *12*(4), e0176322. https://doi.org/10.1371/journal.pone.0176322

Claxton, G. (2008). Cultivating positive learning dispositions. In *The Routledge Companion to Education*. Routledge.

Claxton, G. (2012). *School as an epistemic apprenticeship: The case of building learning power. Volume 32 of Vernon-Wall Lecture*. https://www.education.sa.gov.au/sites/default/files/school_as_an_epistemic_apprenticeship.pdf?acsf_files_redirect

Claxton, G. (2015). *Intelligence in the flesh*. Yale University Press.

Claxton, G. (2018). *The learning power approach: Teaching learners to teach themselves*. Crown House Publishing.

Costa, A. L., & Kallick, B. (Eds.). (2008). *Learning and leading with habits of mind: 16 essential characteristics for success*. Association for Supervision and Curriculum Development.

Cowie, B., Hipkins, R., Keown, P., & Boyd, S. (2011). *The shape of curriculum change: A short discussion of key findings from the Curriculum Implementation Studies (CIES) project*. New Zealand Council for Educational Research and University of Waikato.

Curwen, M. S., Ardell, A., MacGillivray, L., & Lambert, R. (2018). Systems thinking in a second grade curriculum: Students engaged to address a statewide drought. *Frontiers in Education*, *3*, 90. https://doi.org/10.3389/feduc.2018.00090

Damelin, D., Krajcik, J., McIntyre, C., & Bielik, T. (2017, January). Students making systems models: An accessible approach. *Science Scope*, *40*(5), 78–82. https://doi.org/10.2505/4/ss17_040_05_78

Damelin, D., Stephens, L., & Shin, N. (2019, Fall). Engaging in computational thinking through system modeling. *@Concord, 23*(2), 4–6.

Dauer, J., & Dauer, J. (2016). A framework for understanding the characteristics of complexity in biology. *International Journal of STEM Education, 3*(13), 1–8. https://doi.org/10.1186/s40594-016-0047-y

Davis, B., Sumara, D., & Luce-Kapler, R. (2015). *Engaging minds: Cultures of education and practices of teaching.* Routledge. https://doi.org/10.4324/9781315695891

Deakin Crick, R. (2014). Learning to learn: A complex systems perspective. In R. Deakin Crick, C. Stringher, & K. Ren (Eds), *Learning to Learn: International Perspectives from Theory and Practice.* Routledge. https://doi.org/10.4324/9780203078044

Deloitte, (n.d.). *Impact of the Maker Movement* [Report of Maker Impact Summit, December 2013]. Deloitte Center for the Edge and Maker Media.

DeVane, B., Durga, S., & Squire, K. (2010). "Economists who think like ecologists": Reframing systems thinking in games for learning. *E-Learning and Digital Media, 7*(1), 3–20. https://doi.org/10.2304/elea.2010.7.1.3

Dorsey, C. (2020, Spring). Environments for coherent, inquiry-based learning. *@Concord, 24*(1), 2–3.

Drake, J., Kupers, R., & Hipkins, R. (2017, Spring/Autumn). Complexity—A big idea for education? *IS International School, 19*(2), 30–33.

Duschl, R. A. (2019). Learning progressions: Framing and designing coherent sequences for STEM education. *Disciplinary and Interdisciplinary Science Education Research, 1*(1), 4. https://doi.org/10.1186/s43031-019-0005-x

Duschl, R. A., Maeng, S., & Sezen, A. (2011). Learning progressions and teaching sequences: A review and analysis. *Studies in Science Education, 47*(2), 123–182. https://doi.org/10.1080/03057267.2011.604476

Eames, C., Ritchie, J., Birdsall, S., & Milligan, A. (2020). Climate Change: Prepare Today, Live Well Tomorrow—A review. *Set: Research Information for Teachers, 3*, 40–47. https://doi.org/10.18296/set.0185

Egan, K. (2010). *Learning in Depth: A simple innovation that can transform schooling.* University of Chicago Press. https://doi.org/10.7208/chicago/9780226190457.001.0001

Epstein, J. (2008). Why model? *Journal of Artificial Societies and Social Simulation, 11*(4), 12.

Gerber, J. (2017, December 4). How to use the "iceberg" to understand complex systems. *Agricultural Systems Thinking.* https://agsystemsthinking.net/2017/11/13/iceberg/

Gerwertz, C. (2020, December 2). Teaching math through a social justice lens. *Education Week, 40*, 14–16.

Gilbert, L. A., Gross, D. S., & Kreutz, K. J. (2019). Developing undergraduate students' systems thinking skills with an InTeGrate module. *Journal of Geoscience Education, 67*(1), 34–49. https://doi.org/10.1080/10899995.2018.1529469

Goleman, D., & Senge, P. M. (2014). *The triple focus: A new approach to education* (1st ed). More Than Sound.

Gould, S. J. (1989). *Wonderful life: The Burgess Shale and the nature of history.* Hutchinson Radius.

Hatcher, A., Bartlett, C., Marshall, A., & Marshall, M. (2009). Two-eyed seeing in the classroom environment: Concepts, approaches, and challenges. *Canadian Journal of Science, Mathematics and Technology Education, 9*(3), 141–153. https://doi.org/10.1080/14926150903118342

Heinrich, S., & Kupers, R. (2019). Complexity as a big idea for secondary education: Evaluating a complex systems curriculum. *Systems Research and Behavioral Science, 36*(1), 100–110. https://doi.org/10.1002/sres.2547

Heke, I., Rees, D., Swinburn, B., Waititi, R. T., & Stewart, A. (2019). Systems thinking and indigenous systems: Native contributions to obesity prevention. *AlterNative: An International Journal of Indigenous Peoples, 15*(1), 22–30. https://doi.org/10.1177/1177180118806383

Hipkins, R. (2019a). *"Platforming" the science curriculum: A strategy for reframing content.* Post-normal science education: what might it look like?, NZARE. https://www.nzare.org.nz/assets/Uploads/Science/Paper-for-SIG-symposium-R.-Hipkins.pdf

Hipkins, R. (2019b). Weaving a local curriculum from a visionary framework document. *European Journal of Curriculum Studies, 5*(1), 742–752.

Hipkins, R., & Arcus, C. (1997). Teaching science in context: Challenges and choices. In *Developing the science curriculum in Aotearoa New Zealand* (pp. 101–113). Addison Wesley Longman New Zealand.

Hipkins, R., Bolstad, R., Boyd, S., & McDowall, S. (2014). *Key competencies for the future.* NZCER Press.

Hipkins, R., & Bull, A. (2015). Science capabilities for a functional understanding of the nature of science. *Curriculum Matters, 11*, 117–133. https://doi.org/10.18296/cm.0007

Hipkins, R., Bull, A., & Joyce, C. (2008). The interplay of context and concepts in primary school children's systems thinking. *Journal of Biological Education, 42*(4), 73–77. https://doi.org/10.1080/00219266.2008.9656114

Hipkins, R., & Vaughan, K. (2019). *Subject choice for the future of work: Insights from research literature*. Productivity Commission; New Zealand Council for Educational Research. https://www.nzcer.org.nz/research/publications/subject-choice-future-work-insights-research-literature

Hmelo-Silver, C. E., Jordan, R., Eberbach, C., & Sinha, S. (2017). Systems learning with a conceptual representation: A quasi-experimental study. *Instructional Science*, *45*(1), 53–72. https://doi.org/10.1007/s11251-016-9392-y

Holbert, N. (2016). The powerful ideas of making: Building beyond the curriculum. *Journal of Innovation and Entrepreneurship*, *5*(1). https://doi.org/10.1186/s13731-016-0058-4

Huang, J., Hmelo-Silver, C. E., Jordan, R., Gray, S., Frensley, T., Newman, G., & Stern, M. J. (2018). Scientific discourse of citizen scientists: Models as a boundary object for collaborative problem solving. *Computers in Human Behavior*, *87*, 480–492. https://doi.org/10.1016/j.chb.2018.04.004

Hughes, T. P., Kerry, J. T., Connolly, S. R., Baird, A. H., Eakin, C. M., Heron, S. F., Hoey, A. S., Hoogenboom, M. O., Jacobson, M., Liu, G., Pratchett, M. S., Skirving, W., & Torda, G. (2019). Ecological memory modifies the cumulative impact of recurrent climate extremes. *Nature Climate Change*, *9*(1), 40–43. https://doi.org/10.1038/s41558-018-0351-2

Hughson, T. A., & Wood, B. E. (2020). The OECD Learning Compass 2030 and the future of disciplinary learning: A Bernsteinian critique. *Journal of Education Policy*, 1–21. https://doi.org/10.1080/02680939.2020.1865573

International Baccalaureate Organization. (2020). *Compassionate Systems Approach*. International Baccalaureate Organization (UK).

Jacobson, M., & Wilenski, U. (2006). Complex systems in education: Scientific and educational importance and implications for the learning sciences. *The Journal of the Learning Sciences*, *15*(1), 11–34. https://doi.org/10.1207/s15327809jls1501_4

Jess, M., Atencio, M., & Carse, N. (2018). Integrating complexity thinking with teacher education practices: A collective yet unpredictable endeavour in physical education? *Sport, Education and Society*, *23*(5), 435–448. https://doi.org/10.1080/13573322.2016.1225195

Johnstone, J. F., Allen, C. D., Franklin, J. F., Frelich, L. E., Harvey, B. J., Higuera, P. E., Mack, M. C., Meentemeyer, R. K., Metz, M. R., Perry, G. L., Schoennagel, T., & Turner, M. G. (2016). Changing disturbance regimes, ecological memory, and forest resilience. *Frontiers in Ecology and the Environment*, *14*(7), 369–378. https://doi.org/10.1002/fee.1311

Judson, G. (2018). *A walking curriculum: Evoking wonder and developing sense of place (K-12)*. Author.

Kahn, S., & Zeidler, D. L. (2019). A conceptual analysis of perspective taking in support of socioscientific reasoning. *Science & Education*, *28*(6–7), 605–638. https://doi.org/10.1007/s11191-019-00044-2

Klopfer, E., Scheintaub, H., Huang, W., Wendel, D., & Roque, R. (2009). The simulation cycle: Combining games, simulations, engineering and science using *StarLogo TNG*. *E-Learning and Digital Media*, *6*(1), 71–96. https://doi.org/10.2304/elea.2009.6.1.71

Kuhn, D. (2020). Why is reconciling divergent views a challenge? *Current Directions in Psychological Science*, *29*(1), 27–32. https://doi.org/10.1177/0963721419885996

Lee, T. D., Gail Jones, M., & Chesnutt, K. (2019). Teaching systems thinking in the context of the water cycle. *Research in Science Education*, *49*(1), 137–172. https://doi.org/10.1007/s11165-017-9613-7

Lemmer, M., Kriek, J., & Erasmus, B. (2020). Analysis of students' conceptions of basic magnetism from a complex systems perspective. *Research in Science Education*, *50*(2), 375–392. https://doi.org/10.1007/s11165-018-9693-z

Levrini, O., & Fantini, P. (2013). Encountering productive forms of complexity in learning modern physics. *Science & Education*, *22*(8), 1895–1910. https://doi.org/10.1007/s11191-013-9587-4

Li, J., & Wilenski, U. (2009). *NetLogo Sugarscape 1 Immediate Growback model*. Center for Connected Learning and Computer-Based Modeling, Northwestern University, Evanston, IL. http://ccl.northwestern.edu/netlogo/models/Sugarscape1ImmediateGrowback.

Lucas, B., & Hanson, J. (2016). Thinking like an engineer: Using engineering habits of mind and signature pedagogies to redesign engineering education. *International Journal of Engineering Pedagogy (IJEP)*, *6*(2), 4. https://doi.org/10.3991/ijep.v6i2.5366

Lux, J.-D., & Budke, A. (2020). Playing with complex systems? The potential to gain geographical system competence through digital gaming. *Education Sciences*, *10*(5), 130. https://doi.org/10.3390/educsci10050130

Marope, M., Griffin, P., & Gallagher, C. (n.d). *Future competences and the future of curriculum: A global reference for curricula transformation*. UNESCO, International Bureau of Education.

McDowall, S., & Hipkins, R. (2019). *Curriculum integration: What is happening in New Zealand schools?* New Zealand Council for Educational Research. https://www.nzcer.org.nz/system/files/Curriculum%20Integration%202018-2019.pdf

McGrew, S., Breakstone, J., Ortega, T., Smith, M., & Wineburg, S. (2018). Can students evaluate online sources? Learning from assessments of civic online reasoning. *Theory & Research in Social Education, 46*(2), 165–193. https://doi.org/10.1080/00933104.2017.1416320

Ministry of Education. (2007). *The New Zealand curriculum*. Learning Media. https://nzcurriculum.tki.org.nz/The-New-Zealand-Curriculum

Montes de Oca Munguia, O., Harmsworth, G., & Young, R. (2009). *The use of an agent-based model to represent Maōri cultural values*. 18th World IMACS ? MODSIM Congress, Cairns, Australia. https://icm.landcareresearch.co.nz/knowledgebase/publications/public/MODSIM09_Agentbased_modelling_.pdf

Nelsen, P. J. (2015). Intelligent dispositions: Dewey, habits and inquiry in teacher education. *Journal of Teacher Education, 66*(1), 86–97. https://doi.org/10.1177/0022487114535267

Newton, M. H., & Zeidler, D. L. (2020). Developing socioscientific perspective taking. *International Journal of Science Education*, 1–18. https://doi.org/10.1080/09500693.2020.1756515

Nordine, J., Fortus, D., Lehavi, Y., Neumann, K., & Krajcik, J. (2018). Modelling energy transfers between systems to support energy knowledge in use. *Studies in Science Education, 54*(2), 177–206. https://doi.org/10.1080/03057267.2018.1598048

OECD. (2019). *OECD Future of Education and Skills 2030: OECD Learning Compass 2030: A seires of concept notes*. OECD. http://www.oecd.org/education/2030-project/teaching-and-learning/learning/learning-compass-2030/OECD_Learning_Compass_2030_Concept_Note_Series.pdf

O'Neill, D. W., Fanning, A. L., Lamb, W. F., & Steinberger, J. K. (2018). A good life for all within planetary boundaries. *Nature Sustainability, 1*(2), 88–95. https://doi.org/10.1038/s41893-018-0021-4

Osler, A. (2017). Teaching for cosmopolitan citizenship. *Educational Leadership, December 2016/ January 2017*, 42–46.

Parsons, M., & Thoms, M. (2021, April 5). Floodplains aren't separate to a river—they're an extension of it. It's time to change how we connect with them. *The Conversation*. https://theconversation.com/floodplains-arent-separate-to-a-river-theyre-an-extension-of-it-its-time-to-change-how-we-connect-with-them-

Penetito, W. (2009). Placed-based education: Catering for curriculum, culture and community. *New Zealand Annual Review of Education, 18*, 5–29.

Perkins, D. (2010). *Making learning whole: How seven principles of teaching can transform education*. Jossey-Bass.

Perkins, D. (2014). *Future wise: Educating our children for a changing world*. Jossey-Bass.

Petersen, J. E., Frantz, C. M., Tincknell, E., & Canning, C. (2018). An animated visual representation of real-time resource flows through a community enhances systems thinking: Systems thinking is enhanced by animation of community resource flows. *Systems Research and Behavioral Science*, *35*(6), 718–737. https://doi.org/10.1002/sres.2514

Phillipson, N., & Wegerif, R. (2020). The Thinking Together approach to dialogic teaching. In *Deeper learning, dialogic learning, and critical thinking* (pp. 32–47). Routledge. https://doi.org/10.4324/9780429323058-3

Pierson, C. M., Anderson, D., & Luczak-Roesch, M. (2020). Developing science capabilities for citizenship through participation in online citizen science (OCS) projects. *Set: Research Information for Teachers*, (1), 19–26. https://doi.org/10.18296/set.0157

Plate, R., & Monroe, M. (2014). A structure for assessing systems thinking. *The Creative Learning Exchange*, *23*(1), 1–6.

Ramírez-Vizcaya, S., & Froese, T. (2019). The enactive approach to habits: New concepts for the cognitive science of bad habits and addiction. *Frontiers in Psychology*, *10*, 301. https://doi.org/10.3389/fpsyg.2019.00301

Raworth, K. (2017). *Doughnut economics: Seven ways to think like a 21st-century economist*. Random House Business Books.

Rayne, A., Byrnes, G., Collier-Robinson, L., Hollows, J., McIntosh, A., Ramsden, M., Rupene, M., Tamati-Elliffe, P., Thoms, C., & Steeves, T. E. (2020). Centring indigenous knowledge systems to re-imagine conservation translocations. *People and Nature*, *2*(3), 512–526. https://doi.org/10.1002/pan3.10126

Rehmat, A., Hmelo-Silver, C. E., Housh, K., Liu, L., & Cisterna, D. (2020). Assessing systems thinking across disciplines: A learning progression. *The Interdisciplinarity of the Learning Sciences*, *2*, 899–900. https://repository.isls.org/handle/1/6820

Reynolds, S. (2020). Recycling as seduction: Critiquing the practice of climate-change education from a primary classroom. *Set: Research Information for Teachers*, *3*, 12–17. https://doi.org/10.18296/set.0181

Robinson, V. (n.d.). *Ladder of inference*. https://www.educationalleaders.govt.nz/Problem-solving/Online-tools-and-resources/Ladder-of-inference

Sacks, O. (1985). *The man who mistook his wife for a hat and other clinical tales*. Gerald Duckworth (UK); Summit Books (USA).

Sacks, O. (2015). *On the move: A life*. Picador. https://doi.org/10.1080/0449010X.2015.1051697

Salado, A., Chowdhury, A. H., & Norton, A. (2019). Systems thinking and mathematical problem solving. *School Science and Mathematics*, *119*(1), 49–58. https://doi.org/10.1111/ssm.12312

Sammel, A. (2020). How embedding indigenous knowledge systems will help the teaching and learning of Western science evolve. In A. Sammel, S. Whatman, and L. Blue (Eds.), *Indigenizing education: Discussions and case studies from Australia and Canada* (pp. 121–144). Springer Nature Singapore. https://doi.org/10.1007/978-981-15-4835-2_6

Samon, S., & Levy, S. T. (2020). Interactions between reasoning about complex systems and conceptual understanding in learning chemistry. *Journal of Research in Science Teaching*, *57*(1), 58–86. https://doi.org/10.1002/tea.21585

Science for Technological Innovation. (2021, March). Veracity Mission Update, March 2021. https://www.sftichallenge.govt.nz/news/veracity-mission-update-aotearoa-new-zealands-opportunity-to-trade-in-trust/

Senge, P. (1990). *The fifth discipline: The art and practice of the learning organization*. Doubleday/Currency.

Shareef, J. (2020, April 29). An indigenous view on doughnut economics from New Zealand. *Moonshot City*. https://www.projectmoonshot.city/

Sheldrake, M. (2020). *Entangled life: How fungi make our worlds, change our minds and shape our futures*. The Bodley Head.

Shepardson, D., Niyogi, D., Roychoudhury, A., & Hirsch, A. (2012). Conceptualizing climate change in the context of a climate system: Implications for climate and environmental education. *Environmental Education Research*, *18*(3), 323–352. https://doi.org/10.1080/13504622.2011.622839

Smith, L., Maxwell, T. K., Puke, H., & Temara, P. (2016). Indigenous knowledge, methodology and mayhem: What is the role of methodology in producing indigenous insights? A discussion from mātauranga Māori. *Knowledge Cultures*, *4*(3), 131–156.

Silva Pacheco, C. (2020). Art education for the development of complex thinking metacompetence: A theoretical approach. *International Journal of Art & Design Education*, *39*(1), 242–254. https://doi.org/10.1111/jade.12261

Soep, E. (2006). Critique: Assessment and the production of learning. *Teachers College Record*, *108*(4), 748–777. https://doi.org/10.1111/j.1467-9620.2006.00667.x

Sommer, C., & Lucken, M. (2010). System competence—Are elementary students able to deal with a biological system? *NorDiNa*, *6*(2), 125–143. https://doi.org/10.5617/nordina.255

Staudt, C., Lee, H.-S., & Roderick, S. (2018, Fall). Solving big problems requires understanding complex systems. *@Concord, 22*(2), 12–13.

Sternberg, R. (2020). Rethinking what we mean by intelligence. *Phi Delta Kappan, 102*(3). https://doi.org/10.1177/0031721720970700

Storey, B., & Butler, J. (2013). Complexity thinking in PE: Game-centred approaches, games as complex adaptive systems, and ecological values. *Physical Education & Sport Pedagogy, 18*(2), 133–149. https://doi.org/10.1080/17408989.2011.649721

Sumara, D. (2002). *Why reading literature in school still matters: Imagination, interpretation and insight*. Lawrence Earlbaum Associates. https://doi.org/10.4324/9781410603449

Tasquier, G., Pongiglione, F., & Levrini, O. (2014). Climate change: An educational proposal integrating the physical and social sciences. *Procedia: Social and Behavioural Sciences, 116*, 820–825. https://doi.org/10.1016/j.sbspro.2014.01.304

The Systems Thinker. (2018). *The Ladder of Inference*. https://thesystemsthinker.com/the-ladder-of-inference/

The Waters Foundation. (n.d.). *The impact of the Systems Thinking in Schools Project: 20 years of research, development and dissemination*. Author. https://waterscenterst.org/wp-content/uploads/2017/07/STIS_Research.pdf

Tolbert, S., Mackey, G., Manning, R., Hayward, B., & Carter, H.-S. (2020). Social agency and ecoliteracy: Seeds of change for teacher education in uncertain climate futures. *Set: Research Information for Teachers, 3*, 54–60. https://doi.org/10.18296/set.0187

Twist, J., & McDowall, S. (2010). *Lifelong Literacy: The integration of key competencies and reading*. New Zealand Council for Educational Research, Cognition Institute. https://www.nzcer.org.nz/research/publications/lifelong-literacy-integration-key-competencies-and-reading

Tytler, R. (2018). Learning progressions from a sociocultural perspective: Response to "co-constructing cultural landscapes for disciplinary learning in and out of school: The next generation science standards and learning progressions in action." *Cultural Studies of Science Education, 13*(2), 599–605. https://doi.org/10.1007/s11422-016-9777-x

Vaughan, K. (2010). The potential of career management competencies for renewed focus and direction in career education. *The New Zealand Annual Review of Education, 20,* 24–51. https://doi.org/10.26686/nzaroe.v0i20.1569

Vaughan, K., Bonne, L., Eyre, J., New Zealand Council for Educational Research, & Ako Aotearoa National Centre for Tertiary Teaching Excellence. (2015). *Knowing practice: Vocational thresholds for GPs, carpenters, engineering *technicians*. http://www.nzcer.org.nz/system/files/Knowing%20Practice%20Final%2022Jan_DIGITAL_1.pdf

Vears, D. F., & D'Abramo, F. (2018). Health, wealth and behavioural change: An exploration of role responsibilities in the wake of epigenetics. *Journal of Community Genetics*, *9*(2), 153–167. https://doi.org/10.1007/s12687-017-0315-7

Verhoeff, R. P., Boersma, K., Knippels, M.-C. P. J., & Gilissen, M. G. R. (2018). The theoretical nature of systems thinking: Perspectives on systems thinking in biology education. *Frontiers in Education*, *3*(Article 40), 1–11. https://doi.org/10.3389/feduc.2018.00040

Weinberger, D. (2012). *Too big to know: Rethinking knowledge now that the facts aren't the facts, experts are everywhere, and the smartest person in the room is the room*. Basic Books.

Weinberger, D. (2019). *Everyday chaos: Technology, complexity, and how we're thriving in a new world of possibility*. Harvard Business Review Press.

Wheaton, B., Waiti, J., Cosgriff, M., & Burrows, L. (2020). Coastal blue space and wellbeing research: Looking beyond western tides. *Leisure Studies*, *39*(1), 83–95. https://doi.org/10.1080/02614367.2019.1640774

Wilenski, U. (1997). *NetLogo GasLab Two Gas model*. Center for Connected Learning and Computer-Based Modeling, Northwestern University, Evanston, IL. https://ccl.northwestern.edu/netlogo/models/GasLabTwoGas

Wilenski, U. (1999). *NetLogo*. https://ccl.northwestern.edu/netlogo/

Wineburg, S., & McGrew, S. (2017). Lateral reading: Reading less and learning more when evaluating digital information. *SSRN Electronic Journal*. https://doi.org/10.2139/ssrn.3048994

Wood, B. E., & Sheehan, M. (2020). Transformative disciplinary learning in history and social studies: Lessons from a high-autonomy curriculum in New Zealand. *The Curriculum Journal*, (3)3, 495–509. https://doi.org/10.1002/curj.87

Yoon, S. A. (2008). An evolutionary approach to harnessing complex systems thinking in the science and technology classroom. *International Journal of Science Education*, *30*(1), 1–32. https://doi.org/10.1080/09500690601101672

Yoon, S. A., Anderson, E., Koehler-Yom, J., Evans, C., Park, M., Sheldon, J., Schoenfeld, I., Wendel, D., Scheintaub, H., & Klopfer, E. (2017). Teaching about complex systems is no simple matter: Building effective professional development for computer-supported complex systems instruction. *Instructional Science*, *45*(1), 99–121. https://doi.org/10.1007/s11251-016-9388-7

Yoon, S. A., Goh, S.-E., & Park, M. (2018). Teaching and learning about complex systems in k–12 science education: A review of empirical studies

1995–2015. *Review of Educational Research*, *88*(2), 285–325. https://doi.org/10.3102/0034654317746090

Yoon, S. A., Goh, S.-E., & Yang, Z. (2019). Toward a learning progression of complex systems understanding. *Complicity: An International Journal of Complexity and Education*, *16*(1), 1–19. https://doi.org/10.29173/cmplct29340

Zohar, A., & Barzilai, S. (2013). A review of research on metacognition in science education: Current and future directions. *Studies in Science Education*, *49*(2), 121–169. https://doi.org/10.1080/03057267.2013.847261

Zolli, A., & Healy, A. M. (2012). *Resilience: Why things bounce back*. Simon & Schuster.

Acknowledgements

Writing this book has been a roller-coaster ride of learning and collaboration. I am grateful to the many people who have helped me along the way. Particular thanks are due to the senior management team at NZCER: Graeme Cosslett (Tumuaki—Director); Heleen Visser (Tumu Rangahau—General Manager Research & Development); and Sheridan McKinley (Kaiwhakahaere Māori—General Manager, Māori). They supported my work and pushed me to go further than I had initially envisaged I could. David Ellis (Kaiwhakatā Pukapuka–Publisher, NZCER Press) and John Huria (Kaietita Matua–Senior Editor) were also supportive and encouraging. The book has benefitted greatly from their expertise. The following members of our NZCER research team critically read various chapters in areas of their expertise and/or directed my attention to useful resources and reports: Mohamed Alansari, Rachel Bolstad, and Sue McDowall. The progressions chapter greatly benefitted from ongoing conversations with my fellow chief researcher, Charles Darr. Conversations with Josie Roberts, editor of NZCER's journal *Set: Research Information for Teachers* helped my considerably when I was struggling with several of the chapter reading guides. I have also drawn inspiration from conversations with ex-colleagues Ally Bull and Karen Vaughan. Our librarian, Rebecca Lythe, helped me source papers at the outset of the project, when all I had in mind was a literature review.

Many other New Zealand-based education scholars and teacher educators/advisers shared ideas, critically read pieces in areas of their expertise, and through rich conversations inspired me to keep going. They include: Bronwyn Wood and Dayle Andersen (Te Herenga Waka, Victoria University of Wellington); Cathy Buntting, Simon Taylor, and Stephen Ross (University of Waikato); Terry Fenn and Sabina Cleary (Ministry of Education). I am also very grateful to Michal Denny, Head of Science at Auckland Girls' Grammar, who made time in her busy schedule to read the whole first draft, leading to the suggestion of chapter reading guides. Cathy Buntting provided an invaluable critique of these chapter reading guides as they took shape. Very special thanks are due to complexity scientist Markus Luczak-Roesch. Our conversations over a period of months pushed me to sharpen my thinking about what I might be taking for granted and he generously wrote the foreword to the book.

Among my international network of friends and scholars, I am most indebted to Jane Drake, previously a curriculum manager in the IBO schools network, and now an independent consultant. Our learning journeys in complexity pedagogy have variously overlapped and diverged across a number of years. Her reading of pivotal chapters sharpened my thinking but most importantly added those nuggets of practical wisdom that I think teachers will really value. Through her, the book has also benefitted from access to the creativity of others, including Jacob Martin, now principal of an international school in Singapore and Sara Heinrich, now working in Sweden, with whom I shared an illuminating conversation about the trial of a complexity unit in India. Others who helped me along the way include Michael Michie (Australia); JC Couture (Canada); Ralph Levinson and Ruth Crick (England). The papers they variously shared inspired and spurred me on.

Special thanks are also due to artist Sarah Slavik for agreeing to the use of one of her wonderful paintings as a cover image. I first saw her work in an exhibition entitled Whakapapa and the Tree of Life. What an apt metaphor for the wondrous complexity of life on our planet!

Index

Aboriginal Australians 94
ACT framework for student
    portfolios 171–73
adaptive intelligence 167–69
additive factors 42
aesthetic dimensions of learning 209, 211
agent-based models *see* bottom-up models
agents *see* components (agents, variables)
    of systems
Aitken, Viv, *Real in all the ways that
    matter* 80
Anthropocene 221
anthropomorphism 44, 126
aquarium e-learning model 50–51, 52
    Netlogo simulation of aquarium
    system 69–70
arts, and complex-thinking
    capabilities 131–32, 209
assessment 158–59
    see also evidence for assessment;
    progression
    drawings as an assessment
    strategy 160–63
    of gains in knowledge of complex
    systems 160–63
    problem-based tasks 167–69
    purposes 173
Assessment Resource Bank (ARB) 197
    "it depends" thinking 39, 40–41,
    137, 166, 211
    Waterways drawing task 161–62,
    163, 197
    Where Did the Water Go? 28–29
Attenborough, David 221
atua matua 96, 98

beliefs 53, 57, 159, 170
    see also ontological beliefs
    epistemological beliefs 42, 43, 177
    interplay between beliefs and
    actions 114–15
bicultural curriculum 210
binary thinking *see* either/or thinking;
    mind–body binary
Boasa-Dean, Teina 99–100
both/and pairs 79, 90, 134–36, 207, 220
bottom-up models 49–50, 55, 62, 68,
    74, 102
    agent-based computer
    simulations 69–70
    agent-based modelling inquiry
    examples 71
    participatory simulation of agents
    following simple rules 68–69
Boucher, Matt, article about feedback
    loops 30
boundaries of a system 27, 28–29, 196
brain, as a complex adaptive
    system 193–95, 201
Brown, Shae 203–06
building learning power approach and
    resources 115

Cambridge Analytica scandal 214
carbon cycle 196
career opportunities *see* employment
cascading connections 163, 164–65
cascading effects 31
causal-loop diagrams
    COVID-19 17–20, 63, 65
    mātauranga Māori 98–99
    Tolaga Bay waka ama coaching
    problem 127
causality 10, 12, 41, 42, 170, 172, 180,
    184
    causal extent of connections 165–67
    teachers' causal reasoning 43–45
cause-and-effect thinking and
    relationships 10, 22, 38, 41–43, 46,
    87, 113, 131, 163

see also choices; consequences of personal actions
    circular nature 112
    climate change 196
    time delays 112
change-over-time graphs 32, 115
chaos theory 31
choices 107
    see also consumption choices
    barriers to personal action 11–12, 83
    consequences 9–11, 12, 77, 82–83, 93, 112, 134, 146
citizen-science projects 148, 149, 150–52, 154, 176, 186–87, 214
citizenship
    importance of complex systems thinking 3–4, 21, 123, 126, 179, 184, 186, 208–09, 217
    learning experiences for informed citizenship 142
    online civic participation 214–15
    as a purpose for learning 124
    systems literacy 188
cityscape assessment exercise 15, 160
climate as a complex system 196
climate change 4, 85, 101, 118, 145
    barriers to personal action 11–12, 83
    collaborative action 150
    EN-ROADS climate simulator 84
    event happening a second time in rapid succession 33
    indigenous perspectives 94
    interconnections 107–08
    long timescale 196
    models of predicted sea level changes 87
Climate Change and Education for a Sustainable Future 150
Climate Interactive 83–84
clockwork systems *see* complicated systems
closed systems 28–29

CMP systems learning framework 36, 49, 63, 162, 163, 181, 199
    adding contextual knowledge 183–84
    using to support student inquiries 50–51
cognition 192–93
    see also learning; metacognition
collaborative learning 53, 54–55, 132, 133
    assessments 159, 160, 173
    collaboration between schools 137–38
    curriculum design 127, 137–38
    drawings done by a group 160
    situated nature 55–56
    students' integral community engagement 147–50
collective action 3, 11, 12, 123, 124, 150, 184, 212
Common Core Standards for science, United States 178
communities of practice 111
community engagement 147–50
compassionate systems thinking 77–79, 87, 95, 114–15, 123, 132, 134, 143, 145, 167
complex systems
    see also dynamic behaviour of complex systems; non-linearity; sensing
    associated beliefs 43–45
    boundaries 27, 28–28, 196
    development of the field 25–26
    differences from complicated systems 16, 20
    humans situated within 10, 76–77, 82–87, 93
    importance of mathematics in research 14
    key features 10
    lack of consensus about definition 24–25
    states 33–34
    structural features 26–29, 170
    summary of concepts 26–35

synonyms 24–25
systems that learn 32
complex systems thinking
see also compassionate systems thinking; dispositions for complex systems thinking; subjects, contribution to complex systems thinking; thinking habits
bringing thinking and sensing together 79–82
contribution of curriculum subjects 126–34
developing capabilities 177
developmental sequence 179–81
habit formation 111–12
importance for citizenship 3–4, 21, 123, 126, 179, 184, 186, 208–09, 217
importance for teachers 209–11
introducing and integrating into curriculum 16–17
learning focus research study 14–16
in management strategies 211–12
purposes 184
relevance for all learners 121–22
as a way of knowing 210
complexity competence 21
complicated systems 29, 179, 209
associated beliefs 43–45
differences from complex systems 16, 20
components (agents, variables) of systems 26, 33, 35, 36, 162, 183
computer games 55–58, 72
computer simulations 32, 49, 51, 55, 61, 63, 64, 69–72, 74, 83–84, 133, 185
concept mapping 49, 57, 63, 160, 163
conceptual learning 14, 21, 79, 126–28, 136, 158, 160, 163, 164, 168, 177, 184
see also CMP systems learning framework
concepts in combination as an indicator of progression 181–82
difficult concepts 200–02

systems approach 198–200
Concord Consortium 72–73
conflicting accounts of events 42
connections *see* interconnections
consequences of personal actions 9–11, 12, 77, 82–86, 93, 112, 134, 146
conspiracy theories 20
constructivism as active meaning-making 195–202
consumption choices 83, 85–86, 146
energy sources and consumption 83–84
water and electricity consumption study 10–11, 12, 83
content knowledge 122, 126–28, 141, 177
see also subjects, contribution to complex systems thinking
"big understandings" 127
complemented by key competencies 126, 128
content reduction 126–27
"just-in-time" 135
contexts for learning 3, 12–13, 14–15, 16, 40, 54–55, 107, 122, 123, 128, 138, 171, 172, 182, 196
adding contextual knowledge to CMP model 183–84
in assessments 159, 163
choosing generative contexts 141–53
global contexts 154
habits 110, 113, 115, 117–19
integral community engagement 147–50
local contexts 84, 91, 105, 154, 163, 179
mix of contexts 153–54
open-ended and values-laden issues 145–47
overall learning focus 154–55
in progressions 177–78
simple practical response to a complex problem 144–45

uncertainties inherent in datasets 150–53
contingency 35, 40–41
coral bleaching, impact of two seasons of warm seas 33
COVID-19 4
 causal-loop diagram to inform policy-making 17–20, 63, 65
 exponential changes 34–35
 unpredictability 35
creative thinking 134
 subject-specific 131–32
critical thinking 41, 118, 131, 134, 177, 185, 213
curriculum 113
 see also *New Zealand Curriculum (NZC)*
 collaborative and co-ordinated design 127, 137–38
 competences and capabilities 124–25, 126
 five messages 121–23
 "hidden curriculum" 12
 integration 122–23, 134–36, 137
 interdisciplinary curriculum 122, 132
 introducing and integrating complex systems teaching 16–17, 78, 126–34
 local curriculum 123, 137, 154
 multidisciplinary approaches 122, 129
 spiral design 203
 stability of core functions of education 125
 transdisciplinary approaches 122–23
 walking curriculum 97–98, 130, 136
cybernetics 25

Darling-Murray river basin 216–17
data collection decisions 88
data reliability 150–53
data, systems thinking orientation 170
decentring 103, 104, 123, 181–82
 indigenous perspectives 91, 93–100

thought experiment 103
decomposition activities 26–27, 51–52, 70
deterministic effects in systems 179, 181
Dewey, John 109
diffusion 182
 an emergent process 32
 particle behaviour at levels of scale 26–27, 52, 126
 randomness 33–34, 44
digital technologies 71, 133, 186–87, 190
direct processes, compared to emergent processes 32, 52
discernment in action 172
discipline-based inquiry practices 50–51, 122–23, 125, 129–31, 132, 135, 142, 216, 217
disinformation 213–16
dispositions for complex systems thinking 107–08, 122, 136, 141, 152, 159, 169–70, 171, 206
 as "bundles of habits" 107, 109
 fixed-dispositions argument 109
 novel situations 117
 relationship between dispositions and habits 108–09
divergent views 42
diversity of agents in complex systems 33, 34
doughnut economics metaphor 85, 96
 Māori perspective 99–101
drawings
 as an assessment strategy 160–63
 pencil-and-paper drawings 49, 61, 63, 65, 70
Dress a Girl Around the World initiative 144–45
dualism 93–94, 182, 200–02
 see also either/or thinking
 mind–body binary 110–11
dynamic behaviour of complex

systems 19, 20, 29–33, 39, 127, 159
as a basis for learning
interactions 56–58
cascading effects 31
developmental progression in students' understanding 179–81, 183, 185, 186, 188
direct and indirect effects 12, 31
levels of scale 27
modelled in e-learning resources 61
dynamical systems theory 25

ecological memory 33
ecological systems
see also two-eyed seeing metaphor
relationship with social systems 84–86, 99–100, 118–19
wild and remote places 95–96
Edelman, Gerald 193, 194
efficacy
group efficacy 11
self-efficacy 11
either/or thinking 89–90, 101, 125, 126, 127, 128
e-learning resources 61–62, 76, 77
see also bottom-up models; computer games; computer simulations; models; top-down models
affordances 74
design principles 72–74
immersion of students inside systems 82–86
"semi-quantitative" features 62–63
Ellen MacArthur Foundation 86
embedded habits 110
embodiment 123, 133
experiences of complexity 65, 69
habits 110
emergent processes 31–32, 56, 68, 80, 180, 181–82, 188, 196
compared to direct processes 32, 52
complex health problems 147

conceptions 197
participatory simulation of agents following simple rules 68–69
reframing problems using emergence 98
thinking 182
empirical models 35–36
employment 208, 209, 211–12, 215–16
enactive habits 110–11
energy
energy sources and consumption 83–84
water and electricity consumption study 10–11, 12, 83
energy transfer 199–200
EN-ROADS climate simulator 84
Entangled Life: How Fungi Make our Worlds, Change our Minds, and Shape our Futures (Sheldrake) 217–20
environment 123, 145
degradation 95
learning experiences 148–50
societal wellbeing and health of the planet 84–86, 99–100, 118–19
epigenetics 146
epistemological beliefs 42, 43, 177
ethical issues 57, 77, 82, 167
evidence for assessment 158, 159, 160, 172, 178
of building systems-thinking habits 169–71
portfolio of evidence 171–73
explicit teaching 35–36
exponential changes 34–35
extended habits 110

far-from-equilibrium conditions 34
feedback cycles 26, 30–31, 184, 196
see also positive-feedback loops
Figure it Out resource series 129, 130
fixes that fail 128
food choices 146, 147

Food Web task 40–41, 197
force fields 199–200
forest fires 33, 34
fractal geometry 129, 130
fungi 217–21

games 80–81
    see also computer games
Games for Learning conference 80–81
GasLab model 71
global citizenship 117, 118, 154
Globe at Night project 151–52
Goleman, Daniel 77
Good Life for All Within Planetary Boundaries 84–85
Gould, Stephen Jay, *Wonderful Life* 202–03, 209
Greater Wellington Regional Council 87
Gut Bugs project, Liggins Institute 220

habits *see* thinking habits
Habits of Mind 108, 117
Habits of Mind approach and resources 116
Heinrich, Sara 16–17
higher order thinking 185
"hockey stick" graphs 35
humans
    assumptions about superiority 218–19
    complex human-environment interface 219–20
    distinct from nature, in Western knowledge 10, 39, 83, 93, 95
    future on the planet 221
    situated within complex systems 5, 7, 10, 74, 76–77, 82–87, 93, 134, 167, 191, 209, 212, 217, 220–21
Hurunui College, roroa conservation project 149
hyper-simplification of complex concepts 39

iceberg thinking tool 113, 114–15
ImaginEd teacher education group, Canada 97
indigenous knowledge systems 10, 45, 91–92, 218
    see also two-eyed seeing metaphor
    complex systems thinking 91–92
    parallels between indigenous knowledge and complexity 92–93, 96–97
    pattern thought 97, 102, 105, 163
    place-based learning 94–98, 101, 147
    spiralling sense of time 206
    webs of relationships 93–101
industrial-age schooling 10, 38
initial conditions 27, 30
inputs/outputs of systems 27
inquiry approaches to learning 50
"inside systems" perspective 5, 7, 10, 74, 76–77, 82–87, 93, 134, 167, 191, 209, 212, 217, 220–21
interactions in a system 29–30, 32, 36, 180
    interactive factors 42
    pedagogical strategies 57–58
    social interactions 82
interconnections 3–4, 15, 16, 26, 33, 73, 77, 78, 80, 107, 118, 124, 136, 183, 184, 193, 198, 221
    assessment strategies 163–67, 172
    causal extent 165–67, 180
    COVID-19 pandemic 18–20
    hubs 126–27
    Māori 94–96
    materials and products 133–34
International Baccalaureate Organisation (IBO) schools 17, 54, 77, 114, 115, 122
internet 213–16
"it depends" thinking 39, 40–41, 137, 166, 211

kaitiakitanga 94, 118
Key Competencies for the Future (Hipkins et al.) 3–4
Kids Restore the Kepler (KRTK) programme 148
knock-on-effects 172
knowledge
    see also content knowledge; indigenous knowledge systems; mātauranga Māori; Western knowledge system
    knowledge structures 198
    multiplist views 42
    practical knowledge 216
    unthinkable knowledge 216–21
Kupers, Roland 16–17

ladder of influence 53–54, 115, 171
Landcare scientists, collaboration with Māori 102
lateral reading 213–14
learning 79, 115
    see also cognition; collaborative learning; conceptual learning; contexts for learning; e-learning resources; progression
    aesthetic dimensions 209, 211
    bringing thinking and sensing together 79–82
    building learning power approach and resources 115
    as a complex system 56, 176, 204–05, 210
    complex system dynamics as learning interactions 56–58
    inquiry approaches 50
    non-linearity 176
    place-based learning 94–98, 101, 117–18, 147
    sociocultural views 111
Learning in Depth initiative 135–36, 138
LENScience 145–47

levels of scale in systems 26–27, 51–52, 180, 181, 185
lichens 219–20
Lifelong Literacy project 128
linearity 16, 20, 38–39, 40–43, 179, 195, 196, 204, 206, 217
Logic of Life YouTube video 71
Luczak-Roesch, Markus 14, 150–51, 153, 214–15

magnetism 197–99
maker movement 133–34
management strategies, complexity thinking 211–12
Maniapoto Freshwater Cultural Assessment Framework 102
Mantle of the Expert 80
Manurewa High School Business Academy 211–12
Massachusetts Institute of Technology (MIT) 72, 83
mātauranga Māori 91–92, 142, 210
    see also two-eyed seeing metaphor
    causal loop diagram 98–99
    Landcare water quality modelling collaboration 102
    mauri 94
    perspectives on the doughnut model 99–101
    place-based learning 94–96, 147
    species under-represented in Western conservation management 102
    traditional knowledge and biocultural products 215
    whakapapa 93–94
mathematics
    fractal geometry 129, 130
    importance in research of complex systems 14
    non-linear mathematics 62
    social justice approach 153–54
    solving word problems 12–14, 25–26, 35, 40

mauri 94
meaning-making
  about magnetism 197–98
  awareness of changes in our own meaning-making 202–03
  complexities 188–89, 200–02
  constructivism as active meaning-making 195–202
mechanisms 36, 51, 163
  teaching of mechanisms/processes 51–52, 199
memory 110–11
mental models 43–44, 88, 113, 114, 182, 183, 184, 185, 189, 197–98, 199
metacognition 43, 129, 159, 160, 165, 198, 202
  see also cognition
  collective metacognition 55–56
  ladder of influence visual tool 53–54
  metacognitive processing 52–53
  progression 177
  situated nature of shared conversations 55–56
  in student portfolios 172
Mi'kmaq people, Canada 94, 96, 101, 102
mind–body binary 110–11
misconceptions 195–97, 199, 202
  structured 198
models 49, 86, 129, 133, 172, 214
  see also bottom-up models; CMP systems learning framework; computer simulations; top-down models
  critical thinking 177
  deciding what data to collect 88
  explicit assumptions 88
  exploring trade-offs and efficiencies 87
  learning progressions 176–77, 178–86, 187–88
  mental models 43–44, 88, 113, 114, 182, 183, 184, 185, 189, 197–98, 199
  prompting question asking 86–87, 88, 129
  using as an enquiry process 50
multiple connections 163, 164–65, 180
multiplist views of the nature of knowledge 42

negative-feedback loops 30
NetLogo modelling tool 52, 69–70, 71
  GasLab model 71
  student-built simulations 72
  SugarScape simulation 71
neural Darwinism (neural selection) 193, 194
New Zealand Curriculum (*NZC*) 119, 186
  bicultural curriculum 210
  design for complex systems thinking 124–25, 127
  Health and Physical Education learning area 98
  key competencies 3, 4, 67, 108–09, 124, 125, 126, 128, 137, 154
  purpose statements 124
  relating to others key competency 78
  values statements 124
non-communicable diseases (NCDs) 145–46, 147
non-linearity 12, 16, 20, 22, 25, 33, 34–35, 41, 62, 131, 154, 176, 179, 180, 188, 193, 194, 203, 206, 218
  see also complex systems

Oberlin 9, 10–11, 12
obesity 146, 220
  as an emergent phenomenon 98, 100, 147
ocean, appreciation through diving and surfing 95–96
OECD, "learning compass" 124
ontological beliefs 44–45, 194
  see also indigenous knowledge

systems; Western knowledge system
open systems 27, 28–29

part/whole thinking 220
    assessing drawings 162–63
    student progression 175
    teaching strategies 48–50, 55, 122, 134, 172, 198
pattern thought and seeking 97, 102, 105, 117, 129–31, 163, 170, 178
pencil-and-paper drawings 49, 61, 63, 65, 70
perspective-taking 78–79, 81–82, 123, 124, 133, 136, 142, 143, 156, 172, 173, 201, 210
perturbations 30, 33, 34
    The wolves of Yellowstone Park 31, 82
phase shifts 32, 34
phenomena 36
Picton Kindergarten, "cultural locatedness" 149
place-based learning 94–98, 101, 117–18, 147
policies, incorporating systems thinking 17–20, 63, 65, 185, 186
positive-feedback loops 18–19, 30, 70
problem-solving 26, 118
    expectation of quick, correct answers 39–40
    problem-based assessment tasks 167–69
processes
    see also direct processes; emergent processes
    teaching of mechanisms/ processes 51–52
progression 175–76
    across the curriculum 190
    a complex phenomenon 176–78
    complexities of meaning-making 188–89
    concepts in combination as an indicator 181–82
    development of complex thinking capabilities 177
    developmental sequence 179–81
    models 176–77, 178–86, 187–88
    practical strategy for noticing and documenting progress 186–88
    support for assessment 177
    what students can do with their growing knowledge 184–86
Project GUTS (Growing Up Thinking Scientifically) 72
Pūtātara website 117–19

quantum physics 39, 200–02, 204
question asking 13, 14, 15, 40, 86–87, 88, 129, 166, 168, 172

randomness 33–34, 44
    see also stochastic behaviours
recycling 145
reductionist thinking and teaching 38–39, 45, 48–49, 50, 95, 122, 134, 158
refugees
    different realities 79
    flows simulation 65–66
resilience 34

Sacks, Oliver, *A New Vision of the Mind* 194, 201, 202, 203
SageModeler 64
scaling effects in systems 26–27, 51–52, 180, 181, 185
School Journal, May 2020, article on feedback loops 30
school-to-work transition 211–12
self-organisation 25, 31–32, 56
self-reflection 170
Senge, Peter 77
sensing 78–79, 134, 144–45
    bringing thinking and sensing together 79–82, 132
    natural systems 95–96

simplification of complex concepts 39, 200
simulations
    computer simulations 32, 49, 51, 55, 61, 63, 64, 69–72, 74, 83–84, 133, 185
        participatory simulation 68–69
        physical simulations 65–66, 86
single-level thinking 182
situated systems thinking 55
social fields 79
social interactions 82
social justice 142, 143, 153–54
social-science pedagogies 81–82
societal wellbeing and health of the planet 84–86, 99–100, 118–19
split-screen thinking metaphor 115
StarLogo modelling tool 72
static structures and events 44, 45
statistical inquiries 152–53
STEM subjects 20
Sternberg, Robert 167–69
stochastic behaviours 33–34, 44, 181–82
    see also randomness
stock and flow, differentiation 185
stock-and-flow diagrams 27, 49, 64–65, 113, 115, 199–200
    physical simulation example 65–66
    Water Cycle game-like resource update 67–68
subjectivity 170, 171
subjects, contribution to complex systems thinking 126–36
    creative thinking 131–32
    "practical" subjects 132–34
SugarScape simulation 71
sustainability 34, 84, 85, 119, 123, 143
systems
    see also complex systems
    structural features 26–29
systems citizens 21

systems drawings *see* drawings
systems intelligence 78
systems literacy 21, 188, 217
    skills 184–86, 215
systems memory effects 32–33
systems theory
    dynamical 25 (*see also* dynamic behaviour of complex systems)
    general 25–26
systems thinking
    see also compassionate systems thinking; complex systems thinking; thinking habits
    description 9–10
    habits of systems thinkers 112–15
    illustrated card sets, "Habits of a systems thinker" 113–15
    orientations 170–71
    The Triple Focus 77
    word problems in mathematics study 12–14, 25–26, 35, 40

Te Pūnaha Matatini 216
teachers, importance of complexity thinking 209–11
Teachers with Guts 72
Teaching and Learning Research Initiative (TLRI) projects 142–43, 150–51, 186–87
teleological thinking 44
theoretical models 35–36
thinking
    see also both/and pairs; cause-and-effect thinking and relationships; compassionate systems thinking; complex systems thinking; creative thinking; critical thinking; either/or thinking; "it depends" thinking; part/whole thinking; reductionist thinking and teaching; systems thinking
    higher order 185
    single-level, dual-level and

emergent 182
thinking habits 107–08
    4Es 110–11
    building habits for complex systems thinking 111–12, 137
    complex nature of habits 109–11, 194
    evidence of building systems-thinking habits 169–71
    habits of systems thinkers 112–15
    Habits of Systems Thinkers cards 113–15
    Pūtātara website 117–19
    relationship between dispositions and habits 108–09
    resources (other than Habits of Systems Thinkers cards) 115–17
three little pigs, systems thinking exercise 128
threshold experiences 194–95
time as a complex concept 204–06
timescales 12, 94, 196, 206
top-down models 49–50, 55, 62, 63–66, 70, 74
Tōtaranui 250 Trust 148
trade-offs and efficiencies 87
tragedy of the commons 127–28, 201
trauma time 205
trust 19–20
truthfulness 213–16
Tūhoe 99–100
tūrangawaewae 118
two-eyed seeing metaphor 100–01, 147
    use by scientists to connect knowledge systems 101–02

uncertainty 14, 19, 64, 81, 116–17, 123, 132, 151–53, 159, 181, 196, 204, 206, 211
unpredictability 35
unthinkable knowledge 216–21

variables *see* components (agents, variables) of systems
variation in systems 57
Veracity Mission, Science for Technological Innovation 215
visual models 49, 53–54, 121–22
vocational subjects 32–34

walking curriculum 97–98, 130, 136
water and electricity consumption study 10–11, 12, 83
water cycle as a system 27, 28–29
    Water Cycle game-like resource update 67–68
Waters Foundation 88, 112–13
    Thinking Tools Studio 113, 115
Waterways drawing task (ARB) 161–62, 163, 197
webs of relationships 93–100
Western knowledge system 10, 39, 83, 91–92, 95, 116–17, 218–19
    see also two-eyed seeing metaphor
    beliefs about how the world is 43–45, 52, 76, 97
    dualistic thinking 93–94
    habits of Western thinking 109
    mind–body binary 110–11
whakapapa 93–94
whakapuāwai 118–19
wholes and parts *see* part/whole thinking
Wikipedia 214
Wolves of Yellowstone Park, The 31, 82

www.ingramcontent.com/pod-product-compliance
Lightning Source LLC
Chambersburg PA
CBHW051148290426
44108CB00019B/2649